COURSES AND LECTURES - No. 158

ANTONIO RUBERTI
ALBERTO ISIDORI
PAOLO D'ALESSANDRO
UNIVERSITY OF ROME

THEORY OF BILINEAR DYNAMICAL SYSTEMS

**COURSE HELD AT THE DEPARTMENT
FOR AUTOMATION AND INFORMATION
JULY 1972**

UDINE 1972

SPRINGER-VERLAG WIEN GMBH

ISBN 978-3-211-81206-8   ISBN 978-3-7091-2979-1 (eBook)
DOI 10.1007/978-3-7091-2979-1

Note to the Reader.

The following system of numbering and cross-referencing is used in these notes. Each item (definition, theorem, remark etc.) is given a pair of numbers in the left hand margin, the first one indicating the section, the second one is used to number consecutively within each section. When we refer in a section to an item within the same chapter both item numbers are given. For cross references a third number is added on the left to indicate the chapter.

Moreover preceding the text a list of symbols is given.

List of symbols and abbreviations.

$\Rightarrow$	implies
$\Leftarrow$	is implied by
$\Leftrightarrow$	implies and is implied by
$\forall$	for all
r.h.s	right hand side
l.h.s	left hand side
$\triangleq$	is defined to be
$\sim$	is equivalent to
$\triangleleft$	end of discussion
ϵ	is an element of
$\subset$	is a subset
$\subseteq$	is a subset or coincides
$\supset$	contains
$\supseteq$	contains or coincides
$\mathscr{R}(.)$	range space
$\mathscr{N}(.)$	null space
R^n	n dimensional euclidean space
$C[0,T]$	space of continuous functions defined on $[0,T]$
$\lVert . \rVert$	norm
$*$	(super script) coniugate transpose
$\perp$	orthogonal complement
$\oplus$	direct sum
$\otimes$	tensor product
$\square$	special product of matrices defined in section 1.4.

PREFACE

The term *bilinear systems* is usually applied to systems in which the right hand side of the input-state differential equation $\dot{x}(t) = f(x(t), u(t))$ is defined by summing, to a linear function, a bilinear function of the state $x(t)$ and the input $u(t)$; the output of such systems is obtained by means of a linear transformation of the state.

These systems, whose introduction into literature is of fairly recent date, aroused a great deal of interest among researchers, this on account of two principal reasons. The first of these is to be found in the fact that they constitute a rather satisfactory model of many natural and artificial systems, including processes of population generation (from the ones concerning the populations in biological species to the ones concerning neutron population in nuclear fission), many regulation processes at the physiological level (for example the regulation of CO_2 in the respiratory system), many physical processes (such as thermal exchange, chemical reactions) and finally many economic processes (such as control of capital by varying the rate of interest). The second is to be found in the relative ease with which they lend themselves to analytical treatment, thanks to the particular type of non-linearity that characterizes them.

From an analytical point of view, the bilinear systems can be introduced by considering the right hand side $f(x(t), u(t))$ of an arbitrary non-linear differential equation and its possible approximations. The simplest approximation of this function, in fact, is the one that leads to linear systems, and it is interesting to note that the forcing action in this case is additive. Immediately afterwards one would naturally consider the bilinear approximation, obtaining a multiplicative forcing term in addition to the additive one. This multiplicative term can be interpreted as the effect of the linear control applied to the parameters and, therefore, to the structure of the system. For this reason the bilinear systems can be considered as variable structure systems. This characteristic explains the great potential of these systems as models of natural and artificial processes.

Research on bilinear systems has been concerned with system theoretic problems in the strict sense, as well as with problems of modelling and identification, control and optimization. These notes will confine themselves to the former, and

This work was partially supported by Foundazione Ugo BORDONI.

will therefore deal with the analysis of the mathematical descriptions (Chapter I), the structure of the state space (Chapter II), the analysis of the properties of the minimal outline of some specific methods (Chapter IV).

The starting point of these notes is constituted by a paper [1] in which we presented a systematic theory of the realization of bilinear systems. Results obtained in the course of further research have enable us to generalize the theory to the case of bilinear systems with multidimensional input and outputs, and to present it in a unique form that covers both the continuous and the discrete-time case. Sources and references are given in greater detail in the introductions to each of the chapters.

Before concluding this brief preface, it is interesting to note that the theory here developed for bilinear systems is, quite naturally, a generalization of the theory of linear systems. A simple particularization leads back to the familiar results of the linear theory. For the sake of brevity, however, the reader is generally left to perform these particularizations on his own account.

Chapter I

MATHEMATICAL DESCRIPTIONS

1. Introduction

In this chapter we present systematically the mathematical description of bilinear dynamical systems. More precisely we consider both the descriptions through input-state-output equations, which form the basis of the structure theory, and the descriptions through input-output maps, the starting point of the realization theory.

Moreover, a unique formal treatment of continuous and discrete-time systems is made possible thanks to a suitable choice of parameters for describing both kinds of systems.

The chapter is organized as follows. In section 2 the most general bilinear differential equation is introduced and a recursive procedure for computing the solution is given. This solution, through a suitable formalism, can be expressed through Volterra-series expansions. In section 3 the results of a similar analysis for difference equations are presented. Section 4, which is particularly important for the subsequent developments, is devoted to discussing the choice of the most convenient parameters for identifying a given input-state-output description of a bilinear system. Furthermore, a relationship is established between the kernels characterizing the zero-state input-output model for both continuous — and discrete-time systems.

As regards the sources of the material contained in this Chapter, Theorem (2.5) and the procedure for arriving at expression (2.19) were presented in [2]; the remaining part is original.

2. Bilinear differential equations

In this section we present some basic results concerning the equations
(2.1.) Definition — We call "bilinear differential equation" the equation

$$(2.2) \qquad \overset{\circ}{x}(t) = Ax(t) + N[x(t)\otimes u(t)] + Bu(t)$$

where $u(t)\epsilon R^p$, $x(t)\epsilon R^n$. The tensor product $\otimes$ is defined in such a way that $x(t)\otimes u(t)\epsilon R^{np}$, and A,N,B are constant matrices of proper dimensions. ◁

Note that the second term in the r.h.s. of equation (2.2) is the most general bilinear map $R^p \times R^n \to R^n$.

If $u(.)\epsilon C[0,T]$, the solution $x(.)$ of the equation (2.2) on [0,T] with given initial condition $x(0) = \hat{x}$ can usefully be obtained by considering the sequence $\{x_i(.)\}$ of solutions on [0,T] of the equations

$$(2.3) \qquad \dot{x}_o(t) = Ax_o(t) + Bu(t) \qquad x_o(0) = \hat{x}$$

$$(2.4) \qquad \dot{x}_i(t) = Ax_i(t) + N[x_{i-1}(t)\otimes u(t)] + Bu(t)$$

$$x_i(0) = \hat{x} \qquad (i = 1, 2, \ldots)$$

In fact we have the following

(2.5) Theorem — If $u(.)\epsilon C[0,T]$, then

$$(2.6) \qquad \lim_{i\to\infty} x_i(\cdot) = x(\cdot)$$

in the sense of the norm of C[0,T].

Proof — Consider the sequence of functions

$$(2.7) \qquad z_i(t) \overset{\triangle}{=} x(t) - x_i(t) \qquad (i = 0, 1, 2, \ldots)$$

Since

$$x(t) = e^{At}\hat{x} + \int_0^t e^{A(t-\tau)}\{N[x(\tau)\otimes u(\tau)] + Bu(\tau)\}d\tau \qquad (2.8)$$

and

$$x_i(t) = e^{At}\hat{x} + \int_0^t e^{A(t-\tau)}\{N[x_{i-1}(\tau)\otimes u(\tau)] + Bu(\tau)\}d\tau \qquad (i = 1, 2, \ldots) \quad (2.9)$$

then we have

$$z_i(t) = \int_0^t e^{A(t-\tau)}N[z_{i-1}(\tau)\otimes u(\tau)]d\tau \qquad (i = 1, 2, \ldots) \qquad (2.10)$$

Since $u(.)\epsilon C[0,T]$, a constant $M > 0$ exists such that

$$\| e^{A(t-\tau)}N[a\otimes u(\tau)] \| < M \| a \| \qquad \forall t, \tau\epsilon[0,T] \qquad \forall a\epsilon R^n \qquad (2.11)$$

Then, for each $t\epsilon[0,T]$, we have

$$\| z_i(t) \| \leq M \int_0^t \| z_{i-1}(\tau) \| d\tau \qquad (2.12)$$

By subsequent substitution from (2.12) it is easily deduced that

$$\| z_i(t) \| \leq M^i \int_0^t \int_0^{\tau_1} \int_0^{\tau_2} \ldots \int_0^{\tau_{i-1}} \| z_0(\tau_i) \| d\tau_i d\tau_{i-1} \ldots d\tau_1 \qquad (2.13)$$

Since $u(.)\epsilon C[0,T]$, then also $x(.), x_0(.), z_0(.)\epsilon C[0,T]$ and therefore a constant $K > 0$ exists such that

$$\| z_0(t) \| \leq K \qquad \forall t\epsilon[0,T] \qquad (2.14)$$

and from this we conclude

$$\| z_i(t) \| \leq \frac{M^i K}{i!}t^i \leq \frac{(MT)^i}{i!} K \qquad (i = 0, 1, 2, \ldots) \qquad (2.15)$$

Inequality (2.15) shows that the sequence $\{ |z_i(t)| \}$, for each $t\epsilon[0,T]$, has, as an upper bound, the sequence $\{(MT)^i K/i! \}$ which converges to zero for $i \to \infty$. Thus $\{z_i(.)\}$ converges uniformly to zero on $[0,T]$, and $\{x_i(.)\}$ converges uniformly to $x(.)$ on the same interval. ◁

The solution $x(.)$ of the equation (2.2) can therefore be expressed as the limit of the sequence $\{x_i(.)\}$; this latter, on the other hand, can be evaluated as the sum of a series by introducing the sequence of functions

$$(2.16) \qquad\qquad p_0(t) \triangleq x_0(t)$$

$$(2.17) \qquad\qquad p_i(t) \triangleq x_i(t) - x_{i-1}(t) \qquad\qquad (i = 1, 2, \ldots)$$

In fact the solution is given by

$$(2.18) \qquad\qquad x(t) = \sum_{i=0}^{\infty} p_i(t)$$

The function $p_0(t)$ is easily evaluated by solving equation (2.3) while the functions $p_i(t)$ can be obtained from (2.9) and subsequent substitutions. We thus finally obtain

$$(2.19) \qquad x(t) = e^{At}\hat{x} +$$

$$+ \int_0^t e^{At_1} Bu(t-t_1)dt_1 + \sum_{i=2}^{\infty} \int_0^t \ldots \int_0^t \{[e^{At_i}N\{[e^{A(t_{i-1}-t_i)}}$$

$$\cdot \delta_{-1}(t_{i-1}-t_i)N \ldots \{[e^{A(t_1-t_2)}\delta_{-1}(t_1-t_2)Bu(t-t_1)]\otimes$$

$$\otimes u(t-t_2)\} \ldots]\otimes u(t-t_{i-1})\}]\otimes u(t-t_i)\} \, dt_1 \ldots dt_i +$$

$$+ \sum_{i=1}^{\infty} \int_{o}^{t} \cdots \int_{o}^{t} \{[e^{At_i}N\{[e^{A(t_{i-1}-t_i)}\delta_{-1}(t_{i-1}-t_i)N \cdots$$

$$\cdots \{[e^{A(t_1-t_2)}\delta_{-1}(t_1-t_2)N\{e^{A(t-t_1)}\hat{x}\otimes u(t-t_1)\}]\otimes$$

$$\otimes u(t-t_2)\} \cdots]\otimes u(t-t_{i-1})\}]\otimes u(t-t_i)\} \, dt_1 \cdots dt_i . \qquad (2.19)$$

The step functions $\delta_{-1}(t_j - t_{j+1})$, $j = 1, \ldots, i-1$, have been introduced in order to extend to the same value t the upper limits of integration.

At this point we observe that, when used to represent a continuous-time dynamical system, the equation (2.2) is usually associated with the equation

$$y(t) = C x(t) \qquad (2.20)$$

where $y(t)\epsilon R^q$ and C is a constant matrix.

In this case x(t), u(t) and y(t) represent respectively the state, the input and the output at time t.

The expression for the output of this system is easily deduced from equations (2.19) and (2.20).

In order to give an interpretation to the different terms of this output, it is convenient to introduce the

(2.21) Definition – A state θ is a zero state if it is equivalent to the origin of the state-space.

(2.22) Remark – For linear systems this definition coincides with the one given in [3]; otherwise it generally differs. However, the definition (2.21) seems to be more meaningful because it identifies those states whose contribution to the response is identically zero for every admissible input function. For the connection with the concept of unobservability, see Remark (2.39). ◁

Thanks to definition (2.21) we see that the response of this system appears decomposed into the sum of three terms, i.e.

(2.23) $$y(t) = y_{\hat{x}}(t) + y_u(t) + y_{\hat{x}u}(t)$$

of which the first represents the zero-input response, the second represents the zero-state response, and the third depends jointly on the input and the initial state.

The zero-input response has the expression

(2.24) $$y_{\hat{x}}(t) = C e^{At} \hat{x}$$

while $y_u(t)$ and $y_{\hat{x}u}(t)$ may be represented by means of Volterra-series expansions, with symmetrical kernels. To this end, we specialize the rule for constructing the tensor product $\otimes$, of two vectors $a \epsilon R^s$, $b \epsilon R^r$, in the following way

(2.25) $$a \otimes b = \begin{bmatrix} ab_1 \\ \vdots \\ ab_r \end{bmatrix}$$

and then we introduce an operation, denoted by $\Box$, between an $m \times nr$ matrix L and an $n \times s$ matrix M, as defined by (*)

(2.26) $$L \Box M = [L_1 \ldots L_r] \Box M \triangleq [L_1 M \ldots L_r M].$$

The choice (2.25) and the definition (2.26) are such that

(2.27) $$L[(Ma \otimes b)] = (L \Box M) \cdot (a \otimes b)$$

(*) When $r = 1$, the operation $\Box$ coincides with usual matrix product. This case occurs when the input is one-dimensional.

because for each side we have respectively

$$(L_1 \ldots L_r) \cdot \begin{bmatrix} Mab_1 \\ \vdots \\ Mab_r \end{bmatrix} = (L_1 M \ldots L_r M) \cdot \begin{bmatrix} ab_1 \\ \vdots \\ ab_r \end{bmatrix} \qquad (2.28)$$

The equation (2.27) can be used to transform recursively the expression of $y_u(t)$ and $y_{\hat{x}u}(t)$. Referring to the former we have

$$y_u(t) = \int_o^t Ce^{At_1} Bu(t-t_1) dt_1 + \overset{\infty}{\underset{i=2}{\Sigma}} \int_o^t \ldots \int_o^t \{[Ce^{At_i}N] \Box$$

$$\Box \{[e^{A(t_{i-1}-t_i)} \delta_{-1}(t_{i-1}-t_i)N] \Box \{\ldots \Box [e^{A(t_1-t_2)} \delta_{-1}(t_1-t_2)B]\} \ldots\} \cdot \qquad (2.29)$$

$$\cdot [u(t-t_1) \otimes u(t-t_2) \ldots \otimes u(t-t_i)] \cdot dt_1 \ldots dt_i .$$

The parentheses { } denote the order in which the product must be performed. However, in this case — as can be verified through an easy but tedious computation — the order is unessential and therefore the parentheses may be omitted.

Symmetrizing the kernels of (2.29) by summing over all the i ! permutations of the variables $t_1, \ldots, t_i$, we finally arrive at the following expression

$$y_u(t) = \overset{\infty}{\underset{i=1}{\Sigma}} \frac{1}{i!} \int_o^t \ldots \int_o^t W_i(t_1, \ldots, t_i) [u(t-t_1) \otimes \ldots$$

$$\ldots \otimes u(t-t_i)] dt_1 \ldots dt_i \qquad (2.30)$$

where the symmetrical kernels $W_i(t_1, \ldots, t_i)$ are the $q \times p^i$ matrices

$$W_1(t_1) = Ce^{At_1} B \qquad (2.31)$$

$$W_i(t_1, \ldots, t_i) = \sum_{\text{per}} [C\,e^{At_i}N]\,\square\,[e^{A(t_{i-1}-t_i)}\,\delta_{-1}(t_{i-1}-t_i)N]\ldots$$

(2.32)

$$\ldots \square\,[e^{A(t_1-t_2)}\,\delta_{-1}(t_1-t_2)B] \qquad (i > 1)$$

Here $\sum_{\text{per}}$ denotes summation over the i ! permutations of $t_1, \ldots, t_i$.

Similarly, for $y_{\hat{x}u}(t)$ we get the expression

$$y_{\hat{x}u} = \sum_{i=1}^{\infty} \frac{1}{i!} \int_0^t \ldots \int_0^t [Z_i(t_1, \ldots, t_i)\,\square\,e^{A(t-t_1)}]\cdot$$

(2.33)

$$\cdot\,[\hat{x}\otimes u(t-t_1)\otimes u(t-t_i)]\,dt_1\ldots dt_i$$

where the $q \times n^i$ matrices $Z_i(t_1, \ldots, t_i)$ may be formally obtained from (2.31) and (2.32) by replacing the matrix B with an $n \times n$ identity matrix.

3. Bilinear difference equations.

This section is parallel to the preceding one; we start with the

(3.1) Definition — We call a "bilinear difference equation" the equation

(3.2) $$x(k+1) = Ax(k) + N[x(k)\otimes u(k)] + Bu(k)$$

where $u(k)\epsilon R^p$, $x(k)\epsilon R^n$ and A,N,B are constant matrices of proper dimensions. $\lhd$

The solution $x(.)$ could be easily obtained by a recursive technique; it is also possible to adopt a procedure closely parallel to the one used to solve the equation (2.2). This latter approach seems more convenient in order to give to the solution a Volterra-series structure similar to the one obtained for the solution of the differential equation.

For sake of brevity we will here give only the results, for the case in which the equation (3.2) is associated with

$$y(k) = C x(k) \qquad\qquad (3.3)$$

to represent a discrete-time dynamical system; in (3.3), $y(k) \epsilon R^q$ and C is a constant matrix.

If the output response $y(.)$, with initial condition $x(0) = \hat{x}$, is decomposed according to the expression

$$y(k) = y_{\hat{x}}(k) + y_u(k) + y_{\hat{x}u}(k) \qquad\qquad (3.4)$$

the three terms on the r.h.s. have the forms

$$y_{\hat{x}}(k) = C A^k \hat{x} \qquad (k \geq 0) \qquad\qquad (3.5)$$

$$y_u(k) = \sum_{i=1}^{\infty} \frac{1}{i!} \sum_{k_1=0}^{k} \ldots \sum_{k_i=0}^{k} W_i(k_1, \ldots, k_i) \cdot \qquad\qquad (3.6)$$

$$\cdot [u(k-k_1) \otimes \ldots \otimes u(k-k_i)] \qquad (k \geq 0)$$

$$y_{\hat{x}u}(k) = \sum_{i=1}^{\infty} \frac{1}{i!} \sum_{k_1=0}^{k} \ldots \sum_{k_i=0}^{k} [Z_i(k_1, \ldots, k_i) \, N A^{k-k_1}] \cdot \qquad\qquad (3.7)$$

$$\cdot [\hat{x} \otimes u(k-k_1) \otimes \ldots \otimes u(k-k_i)] \qquad (k \geq 0) \;.$$

The symmetrical kernels of the series (3.6) are given by

$$W_1(k_1) = C A^{k_1-1} B \delta_{-1}(k_1-1) \qquad\qquad (3.8)$$

$$W_i(k_1, \ldots, k_i) = \sum_{per} [C A^{k_i-1} \delta_{-1}(k_i-1) N] \Box \ldots \qquad\qquad (3.9a)$$

$$\Box[A^{k_{i-1}-k_i-1}\delta_{-1}(k_{i-1}-k_i-1)N]\Box\ldots$$

(3.9b)

$$\ldots\Box[A^{(k_1-k_2-1)}\delta_{-1}(k_1-k_2-1)B] \qquad (i>1)$$

where the summation is carried out over all the i ! permutations of the variables $k_1,\ldots,k_i$.

The expression for the symmetrical kernels $Z_i(k_1,\ldots,k_i)$ can be formally obtained from the expression (3.9) of the kernels $W_i(k_1,\ldots,k_i)$ by replacing the matrix B by an nxn identity matrix.

(3.10) **Remark** – It is important to stress that the solutions of the equation (3.2) are defined for non negative values of k. Consequently, the expressions (3.8) and (3.9) for the kernels appears affected by the step functions $\delta_{-1}(k_1-1)$ and $\delta_{-1}(k_j-k_{j+1}-1)$, $j=1,\ldots,i-1$, respectively.

From a systemistic point of view, the kernels thus defined satisfy the causality condition. This is not true for the kernels defined by (2.31) and (2.32) in the continuous-time case. This difference corresponds to the well-know one existing in linear systems (which may be considered as the particular case in which N = 0).

(3.11) **Remark** – From the computational point of view it must be stressed that in the series (3.6) and (3.7), for each k, only the first k terms are non zero. This is a consequence of the structure of the kernels. This property immediately appears when equation (3.2) is solved by the recursive technique.

(3.12) **Remark** – It may be useful to note the formal similarity between the structure of the kernels (2.31) and (3.9). Apart from the step functions, the latter may be deduced from the former by replacing e^{At} with A^{k-1}.

4. Bilinear systems.

As was previously pointed out, the bilinear equation (2.2) and the (2.20) can be considered as a description of a continuous-time dynamical systems; this, for convenience, will be called a "bilinear system". Similarly, the equations (3.2) and (3.3) can be considered as a description of a discrete-time bilinear dynamical system. Both kinds of descriptions are characterized by the matrices A,B,C as concerns the linear terms and by the matrix N, together with the rule for constructing the tensor product $\otimes$, as concerns the bilinear term.

If this rule is changed, we get a new matrix $\tilde{N}$ which differs from the previous one in the ordering of the columns. The matrix N corresponding to the rule defined by equation (2.25) is of particular interest; in fact the nxn blocks of the partition of this matrix according to

$$N = [N_1 \ldots N_p] \tag{4.1}$$

appear in the solution of the bilinear equation. This can be observed by computing, for instance, the response of the continuous-time system through the equations (2.24), (2.30) and (2.33), performing the operations $\square$ and keeping in mind that the solution is unique for each fixed u(.) and $\hat{x}$.

If a different rule is chosen for the tensor product, a different matrix $\tilde{N}$ must be considered in the bilinear equation; but in the solution, once again, the columns of $\tilde{N}$ appear grouped together to form exactly the p blocks $N_1, \ldots, N_p$ defined by (4.1).

The special role of the rule (2.25) is thus clear; furthermore we observe that this corresponds to representing the vector $N[x(t) \otimes u(t)]$ as the result of a linear map $R^n \to R^n$, parametrized with u(t), that is

$$N[x(t) \otimes u(t)] = [\sum_{i=1}^{p} N_i u_i(t)] x(t) \tag{4.2}$$

This type of representation will be very useful in the structure analysis of the state space, developed in the following sections.

Keeping in mind this special role of the rule (2.25) and this last observation, we can summarize the previous discussion concluding that the set of matrices

$$(4.3) \qquad\qquad \sigma = (A, N_1, \ldots, N_p, B, C)$$

identifies uniquely the descriptions of both continuous-time and discrete-time bilinear systems.

Bearing in mind the characterization of $x(t)$ (and $x(k)$) as the state variable at time t (and, respectively, k) is natural to associate with a given "system" the set, which will be denoted by Σ, of all the "descriptions" of type (4.3) obtained through a change of variable in the system state space. If T denotes the nxn nonsingular matrix by means of which the coordinate transformation is performed, the set Σ is characterized by

$$(4.4) \qquad \Sigma = \{\sigma : \sigma = (TAT^{-1}, TN_1T^{-1}, \ldots, TN_pT^{-1}, TB, CT^{-1})\}$$

The characterization (4.3) (or (4.4) where necessary), will be used in the subsequent chapters to develop a unique structure analysis of continuous-time and discrete-time bilinear systems.

Later, a realization theory will be developed and the starting point will be the zero-state input-output map of the systems. This map is expressed by (2.30) for continuous-time systems and by (3.6) for discrete-time systems; therefore it is completely characterized by the sequence of kernels (2.31)-(2.32) and, respectively, (3.8)-(3.9). In order to correlate the realization theory for both kind of systems, we give the following

(4.5) **Proposition** — The following relation holds between the derivatives of the kernels of the sequence $\{W_i(t_1, \ldots, t_i)\}_1^\infty$ and the values of the kernels of the sequence

$\{W_i(k_1, \ldots, k_i)\}_1^\infty$

$$\left[\frac{\partial^{h_1-1}}{\partial\theta_1^{h_1-1}} \cdots \frac{\partial^{h_i-1}}{\partial\theta_i^{h_i-1}} W_i(\theta_1+\theta_2+\ldots+\theta_i, \ldots, \theta_{i-1}+\theta_i, \theta_i) \right]_{\substack{\theta_1=0^+ \\ \cdots\cdots\cdots \\ \theta_i=0^+}} = \quad (4.6)$$

$$= W_i(h_1+h_2+\ldots+h_i, \ldots, h_{i-1}+h_i, h_i)$$

$$(h_1, \ldots, h_i = 1, 2, \ldots)$$

Proof – Starting from (2.31), (2.32), (3.8) and (3.9), by direct computation one obtains that both sides of (4.6) are equal to

$$C A^{h_1-1} B \qquad (i=1) \qquad\qquad (4.7)$$

$$\sum_{\text{per}} [C A^{h_i-1} N] \,\square\, [A^{h_{i-1}-1} N] \,\square \ldots \square\, [A^{h_1-1} B] \qquad (i > 1) \qquad (4.8)$$

This completes the proof. ◁

Moreover, the derivatives of the kernels of the sequence $\{W_i(t_1, \ldots ,t_i)\}_1^\infty$ appearing on the l.h.s. of (4.6) uniquely specify the entire sequence, because each term is an analytic function. Consequently, the parameters (4.6) uniquely specify the zero-state input-output map of both the continuous-time and the discrete-time bilinear system.

(4.9) **Remark** – It can easily be checked, by direct substitution, that the zero-state input-output map of a bilinear system remains unchanged under a coordinate transformation in the state space or, what is the same, is invariant on the set Σ (see (4.4)).

Chapter II

STRUCTURE ANALYSIS

1. Introduction

In this chapter we develop a structure analysis of the state-space of bilinear systems, starting from suitable input-state and state-output interaction properties. An interesting feature of this analysis lies in the possibility of generalizing to bilinear systems all the main results which hold in the structure theory of linear systems : state-space decomposability and existence of canonical forms of the equations, dependence of the zero-state input-output map on a subsystem, etc. Another feature to be emphasized is that all these results can be obtained by means of standard linear algebraic tools.

The chapter is organized as follows. Section 2 introduces two important subspaces of the state-space of bilinear systems and gives a procedure for constructing them. These subspaces, which play a key role in the subsequent analysis, suitably generalize the invariant subspaces on which the structure analysis of linear systems is based. In section 3 we define the properties of reachability from the origin and unobservability, and find the least subspace containing all the states which are reachable from the origin, as well as the subspace formed by all the unobservable states. In section 4 we prove the existence of canonical forms of the equation, as well as their uniqueness within a suitable equivalence class.

As concerns the sources of this material, all results given in this chapter were previously established by ourselves, for the case of one-dimensional input and output [1]. The novelty of this chapter consists in the multidimensionality of the input and output.

It is convenient to note that the approach is such as to supply also a procedure for constructing the canonical forms.

An analysis of structure properties of bilinear systems was also given by R.W. Brokett in [4], with a different approach (theory of Lie algebras) ; in particular, in [5], he independently showed the existence of canonical forms of the equations. For other results related to the controllability and reachability of bilinear systems, the reader is referred to [6] [7] [8].

The paper [9] of G. Basile and G. Marro, concerned with the analysis of those invariant subspaces on which the controllability and observability theory of linear systems is based, has been a useful point of reference for the developments given in section 2.

2. Some invariant subspaces

In this section we introduce two suitable subspaces of R^n which are important from the point of view of the properties of reachability and unobservability.

Let $\text{gen}_{A,N_1,\ldots,N_p}(B)$ denote the least subspace of R^n invariant under the matrices $A,N_1,\ldots,N_p$ containing $\Re(B)$. For the construction of this subspace, we introduce the matrix sequences

$$\overline{P}_1 = B$$

$$\overline{P}_i = (A\overline{P}_{i-1} \quad N_1\overline{P}_{i-1} \ldots N_p\overline{P}_{i-1}) \qquad (i = 2, 3, \ldots) \qquad (2.1)$$

$$P_i = (\overline{P}_1 \ldots \overline{P}_i)$$

and have the following

(2.2) Lemma — The least subspace of R^n invariant under $A,N_1,\ldots,N_p$ and containing $\Re(B)$ is

$$\text{gen}_{A,N_1,\ldots,N_p}(B) = \Re(P_n) \qquad (2.3)$$

Proof – If we define the sequence of subspaces

$$\mathcal{P}_1 = \mathfrak{R}(B)$$

(2.4)

$$\mathcal{P}_i = \mathcal{P}_{i-1} + A\mathcal{P}_{i-1} + N_1\mathcal{P}_{i-1} + \ldots + N_p\mathcal{P}_{i-1} \qquad (i = 2, 3, \ldots)$$

we easily verify that

(2.5) $$\mathcal{P}_i = \mathfrak{R}(P_i) \qquad \forall i$$

If, $\mathcal{P}_k = \mathcal{P}_{k-1}$ for some value k, then from (2.4) $\mathcal{P}_{k-1}$ is invariant under $A, N_1, \ldots$ $\ldots, N_p$; moreover $\mathcal{P}_i = \mathcal{P}_{k-1}$ for all $i \geq k$. From this and from

(2.6) $$\mathcal{P}_{i-1} \subseteq \mathcal{P}_i \subseteq R^n$$

which is an immediate consequence of (2.1), it follows that there exists an integer $k_o \leq n$ such that

(2.7) $$\begin{aligned} \mathcal{P}_{i-1} \subset \mathcal{P}_i & \qquad \forall i \leq k_o \\ \mathcal{P}_{i-1} = \mathcal{P}_i & \qquad \forall i > k_o \end{aligned}$$

Therefore $\mathcal{P}_n$ is invariant under $A, N_1, \ldots, N_p$ and, by vertue of (2.1), it contains $\mathfrak{R}(B) = \mathcal{P}_1$.

To prove that $\mathcal{P}_n$ is the least subspace with these properties, observe that any subspace $\mathscr{X}$ of this type must satisfy

(2.8) $$\mathscr{X} \supseteq \mathfrak{R}(B) = \mathcal{P}_1$$

(2.9) $$\mathscr{X} \supseteq A \mathscr{X} \supseteq A\mathcal{P}_1 \; ; \mathscr{X} \supseteq N_1 \mathscr{X} \supseteq N_1\mathcal{P}_1 \; ; \ldots ; \mathscr{X} \supseteq N_p \mathscr{X} \supseteq N_p\mathcal{P}_1$$

and then

(2.10) $$\mathscr{X} \supseteq \mathcal{P}_2 \; .$$

Again, from (2.10)

$$\mathfrak{X} \supseteq A\mathfrak{X} \supseteq A\mathcal{P}_2 \; ; \; \mathfrak{X} \supseteq N_1 \mathfrak{X} \supseteq N_1 \mathcal{P}_2 \; ; \ldots ; \; \mathfrak{X} \supseteq N_p \mathfrak{X} \supseteq N_p \mathcal{P}_2 \qquad (2.11)$$

and then

$$\mathfrak{X} \supseteq \mathcal{P}_3 . \qquad (2.12)$$

Proceeding in this way we have that

$$\mathfrak{X} \supseteq \mathcal{P}_i \qquad \forall i \qquad (2.13)$$

Therefore any subspace of R^n invariant under $A, N_1, \ldots, N_p$ and containing $\mathcal{R}(B)$ contains $\mathcal{P}_n$, which consequently is the least of them. $\triangleleft$

The other subspace to be considered is the largest subspace of R^n invariant under $A, N_1, \ldots, N_p$ contained in $N(C)$. It will be denoted by $\overline{gen}_{A, N_1, \ldots, N_p}(C)$. In order to get the expression of this subspace it is convenient to establish a general result about a sort of duality between the two kinds of invariant subspaces here considered. In this connection we can prove the following

(2.14) Lemma — The largest subspace of R^n invariant under the matrices $A, N_1, \ldots, N_p$ and contained in a given subspace $\mathfrak{X}$ is equal to the orthogonal complement of the least subspace of R^n invariant under $A^*, N_1^*, \ldots, N_p^*$ and containing the orthogonal complement of $\mathfrak{X}$, that is

$$\overline{gen}_{A, N_1, \ldots, N_p}(Z) = [gen_{A^*, N_1^*, \ldots, N_p^*}(Z^*)]^\perp \qquad (2.15)$$

where Z is a matrix such that $\mathcal{N}(Z) = \mathfrak{X}$.

Proof — We start from the property that, for any given subspace $\mathcal{F}$ of R^n and any nxn matrix T,

$$T\mathcal{F} \subseteq \mathcal{F} \Leftrightarrow T^*\mathcal{F}^\perp \subseteq \mathcal{F}^\perp \qquad (2.16)$$

Therefore the r.h.s. of (2.15) is invariant under $A, N_1, \ldots, N_p$. Moreover, since

(2.17) $$\text{gen}_{A^*, N_1^*, \ldots, N_p^*} (Z^*) \supseteq \mathcal{R}(Z^*) = \mathcal{N}^{\perp}(Z)$$

it follows that r.h.s. of (2.15) is contained in $\mathcal{N}(Z) = \mathcal{X}$. Finally, we can prove by contradiction that the r.h.s. of (2.15) is the largest subspace of R^n with these properties. In fact, if there exists a subspace $\mathcal{X}$ of higher dimension, applying the above arguments, it would result that $\mathcal{X}$ is an invariant under $A^*, N_1^*, \ldots, N_p^*$ containing $\mathcal{R}(Z^*)$, of lower dimension than $\text{gen}_{A^*, N_1^*, \ldots, N_p^*}(Z^*)$, and this is a contradiction. ◁

As a consequence of the above Lemmas, introducing the sequences of matrices

$$\overline{Q}_1 = C$$

(2.18) $$\overline{Q}_i = \begin{bmatrix} \overline{Q}_{i-1} A \\ \overline{Q}_{i-1} N_1 \\ \vdots \\ \overline{Q}_{i-1} N_p \end{bmatrix} \quad (i=2,3,\ldots) \quad Q_i = \begin{bmatrix} \overline{Q}_1 \\ \vdots \\ \overline{Q}_i \end{bmatrix}$$

we have the following

(2.19) **Lemma** — The largest subspace of R^n invariant under $A, N_1, \ldots, N_p$ contained in $\mathcal{N}(C)$ is

(2.20) $$\overline{\text{gen}}_{A, N_1, \ldots, N_p}(C) = \mathcal{N}(Q_n)$$

Proof — By (2.15) we have that

(2.21) $$\overline{\text{gen}}_{A, N_1, \ldots, N_p}(C) = [\text{gen}_{A^*, N_1^*, \ldots, N_p^*}(C^*)]^{\perp}.$$

The latter, in turn, by Lemma (2.2) is equal to the subspace $\mathcal{N}(Q_n)$ constructed from the sequence (2.18).

3. Reachability and unobservability

The analysis of the structure of the state space of bilinear systems can be based on the concepts of reachability and unobservability. The results of this analysis will show that in the theory of bilinear systems these concepts play as important a role as in the theory of linear systems.

Due to nonlinearity, it may occur that the sets of states having properties of this kind are not subspaces of the state space. However, subspaces are needed when a state space decomposition is to be performed. Therefore it may be useful to resort to the technique of embedding these sets into suitable subspaces.

From now on we assume that B ≠ 0 in (1.4.4) and we start the structure analysis by introducing the following

(3.1) **Definition** — A state x of the bilinear system (1.4.4) is said to be reachable from the origin if there exists an admissible input function that transfers the origin of the state space into the state x in a finite interval of time (*).

As concerns the states reachable from the origin we have the following :

(3.2) **Theorem** — The subset of all the states of system (1.4.4) reachable from the origin spans a subspace $\mathscr{X}_p$ of R^n which can be expressed as

$$\mathscr{X}_p = gen_{A,N_1,\ldots,N_p}(B) \qquad (3.3)$$

Proof — Assume that (1.4.4) is used to represent a continuous-time system. It is easy to observe that, for any given subspace $\mathscr{X}$ of R^n, one has

$$x(t)\epsilon\,\mathscr{X}\,,\;\;\forall t\epsilon[0,T] \Rightarrow \dot{x}(t)\epsilon\,\mathscr{X}\,,\;\;\;\forall t\epsilon[0,T] \qquad (3.4)$$

By hypothesis, at least a basis $\{x_1,x_2,\ldots,x_r\}$ of $\mathscr{X}_p$ is reachable from the origin. Therefore, for the (3.4),

$$\left[(A + \sum_{i=1}^{p} N_i u_i)x_j + Bu\right]\epsilon\,\mathscr{X}_p \;\;\;\forall u\epsilon R^p \;\;\;(j=1,2,\ldots,r) \qquad (3.5)$$

(*) We observe that, when B=0, the set of states reachable the origin degenerates into $\{0\}$.

But, again for (3.4), also

(3.6) $Bu \in \mathcal{X}_p$ $\forall u \in R^p$

and hence, from this and (3.5)

(3.7) $(A + \sum_{i=1}^{p} N_i u_i) \mathcal{X}_p \subseteq \mathcal{X}_p$ $\forall u \in R^p$

From (3.6) and (3.7) it follows that $\mathcal{X}_p$ contains $\mathcal{R}(B)$ and is invariant under $A, N_1, . \ldots, N_p$.

Therefore $\mathcal{X}_p \supseteq \text{gen}_{A,N_1, \ldots, N_p}(B)$, because the latter by definition is the least subspace with these properties. We now can conclude the proof by showing that $\text{gen}_{A,N_1, \ldots, N_p}(B) \supseteq \mathcal{X}_p$. To this end, by using (3.6) and (3.7), we observe that any subspace of R^n invariant under $A, N_1, \ldots, N_p$ and containing $\mathcal{R}(B)$ is such that in any of its point it is possible to assign only velocities belonging to the subspace itself. Therefore any trajectory starting from the origin cannot leave each of such subspaces ; and this implies that $\text{gen}_{A,N_1, \ldots, N_p}(B)$, the least of them, contains all the trajectories starting from the origin, i.e. contains $\mathcal{X}_p$.

Similarly, (3.3) can be proved in the case of discrete-time systems. ◁

The other property on which the structure analysis is based, is the following

(3.8) Definition − A state x of the system (1.4.4) is unobservable if the component of response depending on the initial state is identically zero for every admissible input function.

(3.9) Remark − It should be stressed that, in the case of linear systems, the unobservable states [3] coincide with the states equivalent to x = 0, i.e. the zero-states. In the present case, we have introduced definitions of zero-state (1.2.21) and of unobservable state (3.8), enjoying the same property. The reason for giving a new, but equivalent, definition, is motivated by the current practice. ◁

As concerns the unobservable states we have the following

(3.10) **Theorem** — The subset of all the unobservable states of the system (1.4.4) is a subspace $\mathcal{X}_q$ of R^n which can be expressed as

$$\mathcal{X}_q = \overline{gen}_{A,N_1,\ldots,N_p}(C) . \tag{3.11}$$

Proof — We again limit the proof to the case of continuous-time systems. From the Definition (3.8) and from (1.2.23), (1.2.24) and (1.2.33), it immediately follows that the set of all the unobservable states in a subspace. Since the component of the response depending on the initial state can be formally considered as a solution of (1.2.2) with $B = 0$, from the (3.4) it follows that

$$(A + \sum_{i=1}^{p} N_i u_i)\, \mathcal{X}_q \subseteq \mathcal{X}_q \qquad \forall u \epsilon R^p \tag{3.12}$$

Moreover, from (1.2.20), we have

$$\mathcal{X}_q \subseteq \mathcal{N}(C) \tag{3.13}$$

Therefore X_q is invariant under $A, N_1, \ldots, N_p$ and is contained in $\mathcal{N}(C)$.

Therefore $\mathcal{X}_q \subseteq gen_{A,N_1,\ldots,N_p}(C)$, because the latter by definition is the largest subspace with these properties. We conclude the proof by showing that $gen_{A,N_1,\ldots,N_p}(C) \subseteq \mathcal{X}_q$. This can be done in the same way as at the end of the proof of Theorem (3.2).◁

It seems useful to conclude this section with the following

(3.14) **Remark** — With a view to Theorems (3.2) and (3.10), Lemma (2.14) establishes a duality relationship between the subspaces $\mathcal{X}_p$ and $\mathcal{X}_q$, that is between reachability from the origin and observability. This provides an extension to bilinear systems of the well-known result of the linear theory.

4. State-space decomposition

Starting from the concepts of reachability from the origin and unobservability or, more precisely, from the subspaces $\mathscr{X}_p$ and $\mathscr{X}_q$ defined in the previous section, it is possible to decompose the state-space $\mathscr{X}$ of the system (1.4.4) into the direct sum of four subspaces $\mathscr{A}, \mathscr{B}, \mathscr{C}, \mathscr{D}$, following the procedure proposed by R.E. Kalman in the case of linear constant systems, i.e.

$$\mathscr{A} = \mathscr{X}_p \cap \mathscr{X}_q$$

$$\mathscr{X}_p = \mathscr{A} \oplus \mathscr{B}$$

(4.1)

$$\mathscr{X}_q = \mathscr{A} \oplus \mathscr{C}$$

$$\mathscr{X} = \mathscr{A} \oplus \mathscr{B} \oplus \mathscr{C} \oplus \mathscr{D}.$$

On the basis of a decomposition of this type, which will be called a canonical decomposition, it is possible to prove the following

(4.2) **Theorem** — Assuming as a basis in the state space of system (1.4.4) the union of bases on the four subspaces $\mathscr{A}, \mathscr{B}, \mathscr{C}, \mathscr{D}$ of a canonical decomposition, the set of matrices of the corresponding description (1.4.3) takes the form (canonical form)

$$\left[\begin{pmatrix} A_{aa} & A_{ab} & A_{ac} & A_{ad} \\ 0 & A_{bb} & 0 & A_{bd} \\ 0 & 0 & A_{cc} & A_{cd} \\ 0 & 0 & 0 & A_{dd} \end{pmatrix}, \begin{pmatrix} N_{aa}^{(i)} & N_{ab}^{(i)} & N_{ac}^{(i)} & N_{ad}^{(i)} \\ 0 & N_{bb}^{(i)} & 0 & N_{bd}^{(i)} \\ 0 & 0 & N_{cc}^{(i)} & N_{cd}^{(i)} \\ 0 & 0 & 0 & N_{dd}^{(i)} \end{pmatrix} \text{ for } i = 1, 2, \ldots, p, \right.$$

(4.3)

$$\left. \begin{pmatrix} B_a \\ B_b \\ 0 \\ 0 \end{pmatrix}, (0, C_b, 0, C_d) \right]$$

Proof — The pattern of the zero submatrices in (4.3) is justified as follows. $\mathcal{X}_q =$
$= \mathcal{A} \oplus \mathcal{B}$ is invariant under $A, N_1, \ldots, N_p$: this implies the zero blocks in the positions (3/1), (3/2), (4/1) and (4/2) of $A, N_1, \ldots, N_p$. $\mathcal{X}_q = \mathcal{A} \oplus \mathcal{C}$ is invariant under
$A, N_1, \ldots, N_p$, and this implies the remaining zero blocks. $\mathcal{X}_p$ contains $\mathcal{R}(B)$ and
hence the blocks in the positions (3) and (4) of B are zero. Similarly $\mathcal{X}_q$ is contained
in $\mathcal{N}(C)$ and this implies the zero blocks of C.

(4.4) Theorem — The set of descriptions of system (1.4.4) which assume the canonical form (4.3) is an equivalence class with respect to the transformations defined by
a nonsingular matrix of the type

$$T = \begin{bmatrix} T_{aa} & T_{ab} & T_{ac} & T_{ad} \\ 0 & T_{bb} & 0 & T_{bd} \\ 0 & 0 & T_{cc} & T_{cd} \\ 0 & 0 & 0 & T_{dd} \end{bmatrix} \tag{4.5}$$

where the partitions are consistent with the dimensions of the subspaces $\mathcal{A}, \mathcal{B}, \mathcal{C}, \mathcal{D}$
respectively.

Proof — It is easy to verify that the transformation defined by (4.5) maintains the
pattern of zero submatrices in all matrices of the set (4.3). Consider now any two descriptions in canonical form ; they correspond to two different choices of subspaces,
$\mathcal{B}', \mathcal{C}', \mathcal{D}'$ and $\mathcal{B}'', \mathcal{C}'', \mathcal{D}''$, in the decomposition (4.1) of the state space. The matrix representing the coordinate transformation between the subspaces corresponding
to these choices must have the pattern of zeros corresponding to (4.5). In fact, since
$\mathcal{A} \oplus \mathcal{B}' = \mathcal{A} \oplus \mathcal{B}''$, the submatrices in the positions (3/1), (3/2), (4/1) and (4/2) must
be zero matrices; similarly, the equality $\mathcal{A} \oplus \mathcal{C}' = \mathcal{A} \oplus \mathcal{C}''$ implies the other zero submatrices.

(4.6) Proposition — The zero-state input-output map of the system (1.4.4) depends only on the "b" part of the canonical decomposition. In other terms, the kernels of this map or the parameters characterizing them ((1.2.31) − (1.2.32) for continuous-type systems, (1.3.8) − (1.3.9) for discrete-time ones or (1.4.6) for both) may be computed by replacing the set of matrices $(A, N_1, \ldots, N_p, B, C)$ with the set of matrices $(A_{bb}, N_{bb}^{(1)}, \ldots, N_{bb}^{(p)}, B_b, C_b)$ of the canonical form (4.3). ◁

This proposition may easily be checked by direct substitution, bearing in mind Remark (1.4.9).

Before concluding this section it is convenient to present a result useful for the explicit computation of the dimension of the subspaces characterizing the canonical decomposition (4.1) of the state space.

(4.7) Theorem − The dimensions of the subspaces defined by (4.1) are

$$n_a = n_p - n_o$$

$$n_b = n_o$$

(4.8)

$$n_c = n_q + n_o - n_p$$

$$n_d = n - n_o - n_q$$

where

$$n_p = \text{rank}(P_n)$$

(4.9) $$n_q = n - \text{rank}(Q_n)$$

$$n_o = \text{rank}(Q_n P_n)$$

Proof − Since, for any two matrices P and Q, the following relation holds

(4.10) $$\text{rank}(QP) = \text{rank}(P) - \dim[\mathcal{R}(P) \cap \mathcal{N}(Q)]$$

from the second of (4.1) it follows that

$$\dim(\mathcal{B}) = \text{rank}\,(Q_n\;P_n) = n_o \qquad (4.11)$$

because $\mathcal{X}_p = \mathcal{R}(P_n)$ and $\mathcal{X}_q = \mathcal{N}(Q_n)$. From (4.11) and the others (4.1) the other relations in (4.8) follow.

Chapter III

MINIMAL REALIZATIONS

1. Introduction

In this chapter we study the properties of the minimal realizations of a given zero-state input-output map. In this case too, we can show results that are conceptually analogous to those holding in the linear theory : equivalence between minimality and structure properties, and uniqueness of minimal realizations. A special aspect of bilinear realizations is the interest attaching to the number of multipliers needed when implementing the realization itself. In this connection we show the remarkable property that when the state space of a bilinear realization has the least dimension, then the realization itself can be implemented with the least number of multipliers.

The chapter is organized as follows. In section 2 we prove a relationship between all bilinear realizations of a given zero-state input-output map; this relation will be the basis of the subsequent proofs. In section 3 we present the main results concerning the properties of the minimal realizations. In section 4 we treat the aforementioned problem of the least number of multipliers.

As concerns the sources of this material, the Theorems here proved generalize the ones already proved by ourselves in [1] for one-dimensional input and output. The approach is different, mainly because the analysis of the properties of the realizations is carried out independently from the realizability conditions. The latter will be examined in the next chapter.

The minimality of bilinear realizations has been also studied, independently, by R.W. Brockett in [5]. He considers an equation of type (2.2) homogeneous in the state (i.e. B=0)and the reachability from an equilibrium state $\chi_0 \neq 0$. Then, he proves, for such realizations, results similar to those expressed by Theorems (3.2) and (3.5).

2. A property of bilinear realizations

Let f denote a given zero-state input-output map and let f_σ denote the zero-state input-output map corresponding to the description σ of a bilinear system. We can state the following :

(2.1) Definition — A set of matrices $\sigma = (A, N_1, \ldots, N_p, B, C)$ is a bilinear realization of a given zero-state input output map f, if $f_\sigma = f$.

From this definition we can establish a property characterizing all the bilinear realizations of a given map f, with the aid of the results previously found for the expressions of the map f_σ.

(2.2) Lemma — For each fixed r and s, the product $Q_r \cdot P_s$ is invariant on the set of all bilinear realizations of a given zero-state input-output map.

Proof — Given any two bilinear realizations σ and $\overline{\sigma}$ of the same map f, it follows from the equality $f_\sigma = f_{\overline{\sigma}}$ and the conclusions inherent in Proposition (1.4.5) that

$$CA^{h_1}B = \overline{C}\,\overline{A}^{h_1}\overline{B} \tag{2.3}$$

$$[CA^{h_i}N] \square [A^{h_{i-1}}N] \square \ldots \square [A^{h_1}B] =$$

$$= [\overline{C}\,\overline{A}^{h_i}\overline{N}] \square [\overline{A}^{h_{i-1}}\overline{N}] \ldots \square [\overline{A}^{h_1}\overline{B}] \qquad (i = 1, 2, \ldots) \tag{2.4}$$

for any integer $h_1, \ldots, h_i$. On the other hand, any element of the product $Q_r \cdot P_s$ coincides with a suitable element of the l.h.s. of (2.3), (2.4). This can be verified by explicitly writing down the product $Q_r \cdot P_s$ and performing the operation $\square$. The same property holds between $Q_r \cdot P_s$, relative to $\overline{\sigma}$, and the r.h.s. of (2.3), (2.4). We therefore have

$$Q_r \cdot P_s = \overline{Q}_r \cdot \overline{P}_s \tag{2.5}$$

and this concludes the proof.

3. Minimality, reachability and observability

In this section we analyse the properties of the minimal bilinear realizations of a given zero-state input-output map. We start with the following

(3.1) **Definition** – A bilinear realization is minimal if the dimension of the state-space is minimal over the set of all possible bilinear realization of the given input-output map.

We can now prove a result concerning the relationship between minimality and the structure properties considered in Chapter 2.

(3.2) **Theorem** – A bilinear realization of a given zero-state input-output map is minimal if and only if the associated state space is spanned by the states reachable from the origin and observable.

Proof – We start with proving, by contradiction, that minimality implies that the state space is spanned by the states reachable from the origin and is observable. In fact, if a realization does not have both these structure properties, then it is possible to find a realization with a lower dimension, by effecting the canonical decomposition (2.4.1) of the state space and noting that the set of matrices $(A_{bb}, N_{bb}^{(1)}, \ldots, N_{bb}^{(p)}, B_b, C_b)$ of the canonical form (2.4.3) is still a realization of the same input-output map (see Proposition (2.4.6)).

The converse may be proved in the following way. Let σ, $\bar{\sigma}$ be any two realizations with their own state spaces spanned by the states reachable from the origin and observable; let n, $\bar{n}$ denote their dimensions. On the basis of Lemma (2.2) and denoting $\max(n, \bar{n})$ with N, we have

(3.3)
$$Q_N \cdot P_N = \bar{Q}_N \cdot \bar{P}_N$$

By Theorem (2.3.2) and Lemma (2.2.2), the n rows of P_N and, respectively, the $\bar{n}$ rows of $\bar{P}_N$ are linearly independent. The same can be said about the n columns of Q_N and the $\bar{n}$ columns of $\bar{Q}_N$, thanks to Theorem (2.3.10) and Lemma (2.2.19). Consequently, the equality (3.3) implies $n = \bar{n}$. Together with the result proved in the first part of this proof, this implies that any realization having the aforementioned structure properties is minimal.

(3.4) **Remark** — Since, by construction, the matrix set $(A_{bb}, N_{bb}^{(1)}, \ldots, N_{bb}^{(p)}, B_b, C_b)$ identified by the canonical decomposition satisfies the conditions of Theorem (3.2), it can be considered as a minimal bilinear realization of the given zero-state input-output map. From this follows the possibility of developing algorithms for reducing any bilinear realization to a minimal one. ◁

> We shall now pass to considering the uniqueness of the minimal realizations.

(3.5) **Theorem** — The minimal bilinear realizations of a given zero-state input-output map are an equivalence class modulo the relation

$$(A, N_1, \ldots, N_p, B, C) \sim (\bar{A}, \bar{N}_1, \ldots \bar{N}_p, \bar{B}, \bar{C}) \Leftrightarrow \begin{aligned} A &= T\,\bar{A}\,T^{-1} \\ N_i &= T\,\bar{N}_i\,T^{-1} \quad (i=1,\ldots,p) \\ B &= T\,\bar{B} \\ C &= \bar{C}\,T^{-1} \end{aligned} \tag{3.6}$$

Proof — The implication ⇐ is trivial (see Remark (1.4.9)). To prove the converse, let $\sigma, \bar{\sigma}$ be any two minimal realizations, and let n denote their dimension. Minimality and Theorem (3.2) imply that $Q_n^* \cdot Q_n$ is nonsingular ; therefore we can define the nxn matrix

$$T = (Q_n^* \cdot Q_n)^{-1} Q_n^* \cdot \bar{Q}_n \tag{3.7}$$

Moreover, by Lemma (2.2), we have

(3.8) $$Q_n \cdot P_n = \bar{Q}_n \cdot \bar{P}_n$$

and therefore

(3.9) $$T^{-1} = \bar{P}_n \cdot P_n^* (P_n \cdot P_n^*)^{-1}$$

Now consider that the columns of AP_n are columns of P_{n+1} and the columns of $N_i P_n$ $(i = 1, \ldots, p)$ are again columns of P_{n+1} that $P_1 = B$ and $Q_1 = C$. By Lemma (2.2) we can write

$$Q_n A P_n = \bar{Q}_n \bar{A} \bar{P}_n$$

$$Q_n N_i P_n = \bar{Q}_n \bar{N}_i \bar{P}_n \qquad (i = 1, 2, \ldots, p)$$

(3.10)

$$Q_n B = \bar{Q}_n \bar{B}$$

$$C P_n = \bar{C} \bar{P}_n$$

from which, bearing in mind the definition (3.7) of T and the expression (3.9) of T^{-1}, we obtain the equivalence relation shown in (3.6).

4. On the number of multipliers

The interest in the realizations of a given input-output map with minimal dimension of the state-space derives also in the possibility of implementing this map with the least number of integrators. However, in the case of bilinear realizations, it is also worth while to consider also how many multipliers are needed to implement the given input-output map. As an example we can consider a bilinear system with one-dimensional input and examine the diagram of Fig. 1. This shows one of the possible ways of simulating a given bilinear realization $\sigma = (A, N, B, C)$ by means of lin-

ear operators and two-inputs multipliers. In this diagram N′ and N″ are such that
N′.N″ = N in particular, if this factorization is such that the inner dimension equals
the rank of N, the number of multipliers needed is the least one necessary for the simulation by means of this type of diagram. Moreover, one can readily verify that no
other diagram can exist which provides a simulation with a lower number of multipliers.

Similar considerations could be used in the case of multi-input bilinear realizations. In this way we can associate to any given bilinear realization σ the least
number of multipliers needed for implementing it, i.e. needed for implementing the
bilinear map $N(x \otimes u)$. This number will be denoted by r_σ.

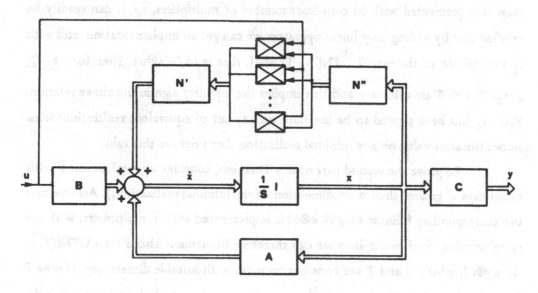

Fig. 1. Simulation of a single-input bilinear system

Starting "aprioristically" in the realization theory, it would therefore seem
interesting to analyze not only the problem of minimizing the state-space dimension,
but also that of minimizing the number r_σ. Actually, however, we shall prove here-
after that state-space minimality implies the minimality of r_σ. More precisely, we

have :

(4.1) Theorem — The number r_σ is invariant on the set of the minimal realizations, where it assumes the least value over the set of all bilinear realizations of a given zero-state input-output map.

Proof — We first prove the invariance of r_σ on any set of realizations equivalent modulo the relations (3.6). For any realization σ, by definition r_σ is the least number of multipliers needed for implementing the bilinear map $N(x \otimes u)$. Consider now any other equivalent realization $\overline{\sigma}$ and let $\overline{x} = Tx$ denote the corresponding state variable; consequently the corresponding bilinear map is $TN[(T^{-1}x) \otimes u]$. If this new bilinear map is implemented with its own least number of multipliers, $r_{\overline{\sigma}}$, it can readily be verified that by adding only linear operators we can get an implementation, still with $r_{\overline{\sigma}}$ multipliers, of the map $T^{-1}TN[T^{-1}Tx) \otimes u]$, that is of $N(x \otimes u)$. Therefore, $r_{\overline{\sigma}} \geq$

$\geq r_\sigma$. But $\sigma, \overline{\sigma}$ are arbitrary and this implies the equality sign in the above relation. Since r_σ has been proved to be invariant on any set of equivalent realizations it assumes the same value on any minimal realization. Let r denote this value.

To prove the second part of this Theorem, consider any realization $\widetilde{\sigma}$ with dimension $\widetilde{n}$ greater than n (= dimension of the minimal realizations). Assume that the corresponding bilinear map $\widetilde{N}(x \otimes u)$ is implemented with $\widetilde{r}$ multipliers. With the same number $\widetilde{r}$ of multipliers we can therefore implement also a map $UT\widetilde{N}[(T^{-1} \cdot U^* x) \otimes u]$, where U and T are constant matrices with suitable dimensions. If now T is a coordinate transformation in the state space such as to induce a canonical decomposition (2.4.3) and if $U = [0 \ I \ 0 \ 0]^*$ is the matrix representing the injection $B \to R^n$, the map $UT\widetilde{N}[(T^{-1}U^* x) \otimes u]$ is the bilinear map of a suitable minimal realization of the same input-output map. This latter, in turn, cannot be implemented with less than r multipliers, and therefore $\widetilde{r} \geq r$. This completes the proof.

Chapter IV

THE CONSTRUCTION OF MINIMAL REALIZATIONS

1. Introduction

In this chapter we consider the problem of constructing minimal realizations. The starting point is the statement of a necessary and sufficient condition for an infinite sequence of Volterra kernels to admit a finite dimensional, constant, bilinear realization. This criterion essentially consists in a sort of "factorizability condition" of the family of kernels. Then we propose two alternative methods for constructing minimal bilinear realizations. Both these methods are based on the assumption that the abovementioned realizability condition is satisfied and operate in two steps : a first one for constructing the "factors" of the given sequence and a second for constructing, from these latter, the minimal realizations.

For the sake of brevity, the treatment refers only to continuous-time systems. However, all the results can easily be proved for discrete-time systems, obviously considering discrete-time functions and, where necessary z transforms instead of the Laplace transforms.

The chapter is organized as follows. In section 2, the condition of realizability is given and a first method of realization is derived. The minimality for the latter is obtained by means of a reduction procedure which is based on the decomposition of state-space and uses results previously established about the minimal realizations. In section 3, a second method of realization is derived, based on some results about the "factorizable" sequences. In this case the minimality is obtained by suitable reducing the starting "factorization". Finally, in section 4, a procedure for the actual construction of a "factorization" of the sequence of kernels is outlined. From this it is proved that, in the case of bilinear systems, a finite number of kernels uniquely specifies all the kernels of the input/output map, and this result is extended to suitable sequences of constant parameters characterizing this map.

The theory of the "factorizable" sequences of kernels, which is developed under more general assumptions than those required by bilinear realizability, is gathered and presented separately in the Appendix.

As regards the sources, the material contained in this chapter is an extension of that presented by the authors in [1] to the multidimensional case, with the addition of Remark (4.10), which is new.

For the sake of completeness, it is also worth noting that another approach is possible to the problem of constructing minimal bilinear realizations from a given nonlinear input/output map. This approach reduces the realization problem to that of matching an infinite sequence of input/output parameters ; on this basis it is possible to develop, for bilinear systems, a realization theory [10] analogous to that originated after a well-known paper of B.L. Ho and R.E. Kalman for linear systems [11].

2. A realizability condition and a first realization procedure

If a zero-state input-output map f can be expanded into a Volterra series with symmetrical kernels $W_i(t_1, \ldots, t_i)$, $i = 1, \ldots, \infty$, with reference to the Definition (3.2.1) we might speak indifferently of realization of the map f or of the family $\{W_i(t_1, \ldots, t_i)\}_1^\infty$. Concerning the realizability conditions we have the following :

(2.1) **Theorem** − A necessary and sufficient condition for a sequence $\{W_i(t_1, \ldots, t_i)\}_1^\infty$ of $q \times p^i$ symmetrical kernels of a Volterra series expansion to be realizable by means of a constant bilinear dynamical system with finite-dimensional state space, is :

a)   that $W_1(t_1)$ has a proper rational Laplace transform

b)   that there exists three matrices $F(t)$, $G(t)$, $H(t)$, respectively $m \times mp$, $m \times p$, $q \times mp$

of functions having proper rational Laplace transform, such that the following relations are satisfied

$$W_i(t_1, \ldots, t_i) = H(t_i) \,\square\, F(t_{i-1} - t_i) \,\square\, \ldots \,\square\, F(t_2 - t_3) \,\square\, G(t_1 - t_2) \tag{2.2}$$

on

$$S_i = \{(t_1, \ldots, t_i) : t_1 > t_2 > \ldots > t_i\} \qquad (i > 1)$$

(2.3) **Remark** — The above condition can also be stated in an alternative way, by referring to any one of the i ! permutations of the variables $(t_1, \ldots, t_i)$ in (2.2). Proof — Necessity. If the sequence $\{W_i(t_1, \ldots, t_i)\}_1^\infty$ is realizable, then, by hypothesis, there exist four matrices A, N, B, C such that the equations (1.2.31) and (1.2.32) are satisfied. The first of these implies the condition a). The equations (1.2.32), considered over the sets S_i defined above, reduces to

$$W_i(t_1, \ldots, t_i) = [Ce^{At_i}N] \,\square\, [e^{A(t_{i-1}-t_i)}N] \,\square\, \ldots \,\square\, [e^{A(t_2 - t_3)}N] \,\square$$
$$\square\, [e^{A(t_1 - t_2)}B] \qquad (i > 1) \tag{2.4}$$

This equation can be read in the form (2.2), by putting

$$Ce^{At}N = H(t) \; ; \; e^{At}N = F(t) \; ; \; e^{At}B = G(t) \tag{2.5}$$

and m = n. Since all these functions have proper rational Laplace transforms, it follows that condition b) is also satisfied.

Sufficiency. Suppose that a) and b) are both satisfied, and consider the matrix

$$L(t) = \begin{bmatrix} W_1(t) & H(t) \\ \\ G(t) & F(t) \end{bmatrix} \tag{2.6}$$

Since all the elements of L(t) have proper rational Laplace transforms, this may be interpreted as the weighting pattern matrix of a constant finite-dimensional linear system with (q + m) outputs and (p + mp) inputs. Consequently there must exist three matrices A, R, S, respectively nxn, nx(p + mp), (q + m)xn, such that

(2.7) $$Se^{At}R = L(t)$$

Let S and R be partitioned in the form

(2.8) $$S = \begin{bmatrix} C \\ \hat{S} \end{bmatrix} \qquad R = (B \ R_1 \ R_2 \ldots R_p)$$

where C is qxn, $\hat{S}$ is mxn, B is nxp, $R_1, R_2, \ldots, R_p$ are nxm; then (2.7) gives

$$W_1(t) = Ce^{At}B$$

$$H(t) = Ce^{At}(R_1 \ R_2 \ldots R_p)$$

(2.9)

$$G(t) = \hat{S}e^{At}B$$

$$F(t) = \hat{S}e^{At}(R_1 \ R_2 \ldots R_p)$$

Now define

(2.10) $$N = (R_1 \ R_2 \ldots R_p) \ \square \ \hat{S}$$

On the basis of this assumption, it can readily be verified that the bilinear system characterized by the matrices A, defined in (2.7), B, C, defined in (2.8) and N verifies the equations (1.2.31) and (1.2.32) on the sets S_i. In fact, substituting (2.9) in (2.2), one has

$$W_i(t_1, \ldots, t_i) = [Ce^{At_i}(R_1 \ldots R_p)] \square \ldots$$

(2.11)

$$\ldots \square[\hat{S}e^{A(t_2 - t_3)}(R_1 \ldots R_p)] \square [\hat{S}e^{A(t_1 - t_2)}B]$$

Performing, for instance, the last operation $\Box$ in (2.11) and taking into account (2.10), we subsequently have

$$[\hat{S}e^{A(t_2 - t_3)}(R_1 \ldots R_p)] \,\Box\, [\hat{S}e^{A(t_1 - t_2)} B] =$$

$$= \hat{S}e^{A(t_2 - t_3)}(R_1 \hat{S}e^{A(t_1 - t_2)} B \ldots R_p \hat{S}e^{A(t_1 - t_2)} B) =$$

$$= [\hat{S}e^{A(t_2 - t_3)}(R_1 \hat{S} \ldots R_p \hat{S})] \,\Box\, [e^{A(t_1 - t_2)} B] = \qquad (2.12)$$

$$= [\hat{S}e^{A(t_2 - t_3)} N] \,\Box\, [e^{A(t_1 - t_2)} B].$$

Proceeding in this way for the other products $\Box$, we obtain (1.2.32), but with the variables $(t_1, \ldots, t_i)$ confined to the sets S_i.

On the other, since $W_i(t_1, \ldots, t_i)$ is symmetrical by definition, the equations (1.2.32) are satisfied for all values of the variable $t_1, \ldots, t_i$. This completes the proof. $\lhd$

Based on the proof of the above theorem and on the results of the analysis developed in the preceding chapters, it is possible to give immediately a first procedure for constructing a minimal bilinear realization of the sequence $\{W_i(t_1, \ldots \ldots, t_i)\}_1^\infty$, once a factorization $\{F(t), G(t), H(t)\}$ of the sequence $\{W_i(t_1, \ldots, t_i)\}_2^\infty$ is available. The corresponding steps are the following :

a) arrange $W_1(t)$, $F(t)$, $G(t)$, $H(t)$ as in (2.6)

b) find a linear realization A, R, S of (2.6)

c) define the bilinear realization A, N, B, C by the equations (2.7), (2.8), (2.10)

d) from this latter derive a minimal realization, on the basis of the results outlined in Remark (3.4) (i.e., by means of the state-space decomposition).

It is convenient to stress that this realization procedure is substantially based on the use of techniques for linear realization and, also that the reduction procedure makes use of standard linear algebraic tools.

3. A second realization procedure

The first realization procedure was based on the possibility of reducing a given realization to the minimal form (i.e. on the connection between state-space decomposition and minimality); the present procedure will be based on the possibility of reducing to its own minimal form a given triplet $\{F(t), G(t), H(t)\}$ associated to a realizable sequence of kernels (see Theorem (2.1)).

In the sequel, such a triplet will be called a **factorization** of the sequence $\{W_i(t_1, \ldots, t_i)\}_2^\infty$ and m its dimension. A factorization will be called **minimal** when its dimension assumes the smallest value over all the factorizations of the sequence.

In the Appendix of this Chapter we present an analysis of the properties of the sequences of kernels satisfying the condition (2.2). In particular we give a condition of minimality, the description of all (minimal and nonminimal) factorizations and a procedure for constructing the minimal ones. Thanks to these results, with reference to the minimal factorization, we can prove the following :

(3.1) **Theorem** – Let $\{W_i(t_1, \ldots, t_i)\}_1^\infty$ be a sequence of bilinearly realizable kernels, and let $\{F_o(t), G_o(t), H_o(t)\}$ be a minimal factorization of the subsequence $\{W_i(t_1, \ldots, t_i)\}_2^\infty$. Then the bilinear realization associated, be means of (2.7), (2.8), (2.10), to a minimal linear realization of

(3.2)
$$\begin{bmatrix} W_1(t) & H_o(t) \\ G_o(t) & F_o(t) \end{bmatrix}$$

is minimal

Proof – Let $F(t)$, $G(t)$, $H(t)$ denote a generic factorization of the given sequence and consider the matrix

(3.3)
$$L(t) = \begin{bmatrix} W_1(t) & H(t) \\ G(t) & F(t) \end{bmatrix}$$

Every linear realization of the matrix L(t) identifies a bilinear realization of the sequence under consideration; viceversa, every bilinear realization can given rise to a factorization and, consequently, a matrix of type (3.3). One can therefore conclude that all the bilinear realizations can be obtained by considering all the linear realizations that can be associated with all the matrices L(t) relating to the given sequence.

In order to identify a bilinear realization having minimum dimension of the state space, one can therefore limit oneself to considering the linear realizations of minimum order for each matrix L(t) and then choosing from among these realizations the one that, in turn, has the minimum order. It follows that

$$n_o = \min_{L(t)} \delta\{L(t)\} \tag{3.4}$$

where $\delta\{L(t)\}$ denote the order of the linear system with weighting pattern L(t).

If in the expression (3.3) for L(t) one now substitutes the expressions for F(t), G(t), H(t) provided by (A.52) and then collects the matrix $\begin{bmatrix} I & 0 \\ 0 & T \end{bmatrix}$ on the left, and its inverse on the right, it becomes easy to see that

$$\delta\{L(t)\} \geq \delta\{L_o(t)\} \qquad \forall L(t) \tag{3.5}$$

where $L_o(t)$ denotes the matrix

$$L_o(t) = \begin{bmatrix} W_1(t) & H_o(t) \\ G_o(t) & F_o(t) \end{bmatrix} \tag{3.6}$$

and this, bearing in mind that $L_o(t)$ is an element of the set of the matrices $L(t)$, concludes the proof. ◁

From this Theorem, there follows a procedure for constructing a minimal bilinear realization :

a)   find a minimal factorization $\{F_o(t), G_o(t), H_o(t)\}$ from the given factorization $\{F(t), G(t), H(t)\}$, following the procedure given in the Appendix.

b)   arrange $W_1(t)$, $F_o(t)$, $G_o(t)$, $H_o(t)$ as in (3.6).

c)   find a minimal linear realization A, R, S of (3.6).

d)   define the bilinear realization A, N, B, C by the equations (2.7), (2.8), (2.10).

We note again that also this second procedure operates through standard linear algebraic methods (reduction of the factorization) and techniques of linear realization.

4. The computation of a triplet F(t), G(t), H(t).

Both the realization procedures considered in the previous sections implicitly require the knowledge of a factorization $\{F(t), G(t), H(t)\}$ of the sequence of kernels under consideration.

It is therefore fundamental to be able to examine if a given sequence of kernels satisfies the realizability condition (Theorem (2.1)), and, in the positive case, to construct a triplet $\{F(t), G(t), H(t)$.

With reference to this second problem, we observe that a procedure for constructing a minimal factorization from a given factorizable sequence of kernels (i.e. satisfying condition (2.2)) follows from the proof of Theorem (A.38) in the Appendix.

For reader's convenience we describe this procedure in the case that p=q=1, although the complexities involved in the general case are only notational.

The underlying assumption is that the sequence of the kernels is factorizable (i.e. a triplet $\{F(t), G(t), H(t)\}$ exist) and that an upper bound m for the dimension m_o of the minimal factorizations is known. In this case the steps are the following :

a)   define the sequence $\{f_i(\theta_1, \ldots, \theta_i)\}_2^\infty$ as

$$f_i(\theta_1, \ldots, \theta_i) = W_i(\theta_1 + \theta_2 + \ldots + \theta_i, \ldots, \theta_1 + \theta_2, \theta_1) \qquad (4.1)$$

and arrange the kernels $f_i(\theta_1, \ldots, \theta_i)$, i = 2, 3, ..., 2m+1 in a matrix

$$S_{rs}(\tau_1, \ldots, \tau_r, \sigma_s, \ldots, \sigma_1) = [\, f_{i+j}(\tau_1, \ldots, \tau_i, \sigma_j, \ldots, \sigma_1) \,] \qquad (4.2)$$

$$i = 1, 2, \ldots, r \qquad\qquad j = 1, 2, \ldots, s$$

(note that, if $\{F(t), G(t), H(t)\}$ is a factorization of the sequence of kernels, then $S_{r,s}(\tau_1, \ldots, \sigma_1) = Q[F, H](\tau_1, \ldots, \tau_r) \cdot P[F, G](\sigma_s, \ldots, \sigma_1)$, as can be verified by particularizing (A.37) and considering explicity the values of the arguments).

b)   find a factorization

$$S_{m,m}(\tau_1, \ldots, \tau_m, \sigma_m, \ldots, \sigma_1) = R(\tau_1, \ldots, \tau_m) \, S(\sigma_1, \ldots, \sigma_m) \qquad (4.3)$$

such that the columns of $R(\tau_1, \ldots, \tau_m)$ and the rows of $S(\sigma_1, \ldots, \sigma_m)$ are linearly independent on the interval $\Delta_m = \{(t_1, \ldots, t_j) : 0 \leqslant t_k \leqslant 1, \forall k\}$.

c)   define a triplet $\{F(t), G(t), H(t)\}$ as follows

$$F(t) = \int_{\Delta_{2m}} U^{-1} R^*(\tau_1, \ldots, \tau_m) \, \hat{S}_{m,m}(\tau_1, \ldots, \tau_m, t, \sigma_m, \ldots, \sigma_1)$$

$$S^*(\sigma_1, \ldots, \sigma_m) V^{-1} d\tau_1 \ldots d\tau_m d\sigma_1 \ldots d\sigma_m \qquad (4.4)$$

where $U^{-1} = \int_{\Delta m} R^* R \, d\tau_1 \ldots d\tau_m$, $V^{-1} = \int_{\Delta m} SS^* d\sigma_1 \ldots d\sigma_m$,

and $\hat{S}_{m,m}(\tau_1, \ldots, \tau_m, t, \sigma_m, \ldots, \sigma_1)$ is a matrix constructed with the last m columns of $S_{m,m+1}(\tau_1, \ldots, \tau_m, t, \sigma_m, \ldots, \sigma_1)$

(4.5) $G(t) = $ first colum of $S(t, \sigma_2, \ldots, \sigma_m)$

(4.6) $H(t) = $ first row of $R(t, \tau_2, \ldots, \tau_m)$

The validity of this procedure may be justified by using the arguments of the proof of Theorem (A.38); in particular, (4.4) is connected with (A.49) and (4.5), (4.6) are connected with (A.45).

(4.7) Remark — The crucial point of this procedure is that of finding the factorization (4.3) for matrix (4.1) ; this, in principle, is a generalization of that occurring in the realization problem of time-varying linear systems, when it is required to factorize the two-variable weighting pattern [12].

(4.8) Remark — For the extension to the general case (p>1, q>1), instead of (4.4) and (4.6), it will be necessary to consider p equations for finding $F^{(i)}(t)$ and, respectively, p equations for finding $H^{(i)}(t)$, i = 1, . . . , p. In these equations there should appear suitable partitions of the matrix corresponding to $S_{m,m+1}$.

Before concluding this section, we observe that from this procedure emerges directly the possibility of proving the following

(4.9) Theorem — A factorizable sequence of kernels $\{W_1(t_1, \ldots, t_i)\}_2^\infty$ is uniquely specified by the sequence $W_i(t_1, \ldots, t_i)_2^{2m_o+1}$, where m_o is the dimension of its minimal factorization.

(4.10) Remark — The importance of this result from the point of view of the modelling of bilinear systems should be clear. We note that this is valid under quite general hypotheses: the only assumption is that the sequence $\{W_i(t_1, \ldots, t_i)\}_2^\infty$ is factorizable.

One may therefore conjecture the existence of a stronger result valid for bilinear systems, since in this case F(t), G(t), H(t) have a proper rational Laplace transform. In effect it is possible to prove that a sequence of kernels realizable by means of a finite dimensional bilinear system can be uniquely specified by assigning a finite number of constant parameters (more precisely : the ones considered in Proposition (4.5)). The proof of this in left to the reader, which may use Lemma (3.2.2), Theorem (3.3.5) and follow the line of the above procedure.

APPENDIX

FACTORIZABLE SEQUENCES OF KERNELS

The condition stated in Theorem (4.2.1), highlights the interest of studying the sequences of kernels factorizable by means of functions with proper rational transforms. Actually, for the analysis of the general properties of these sequences this hypothesis can be dropped out. The sequences thus specified (see Definition (A.1)) are involved in the study of bilinear systems with infinite dimensional state-space [13], and this is the reason why we have presented the relative theory in this appendix.

Before starting the treatment, we would draw the reader's attention to the connection between the matrices here considered (see (A.3) up to (A.6)) and the ones introduced in Chapter 2 (see (2.2.1) and (2.2.18)).

A precise statement of our hypothesis is contained in the following

(A.1) Definition — A sequence $\{W_i(t_1, \ldots, t_i)\}_2$ of qxp^i kernels is said to be factorizable if there exist three analytic functions(*) $F(t)$, $G(t)$, $H(t)$, respectively, mxm, mxp, qxm, such that

$$(A.2) \quad W_i(t_1, \ldots, t_i) = H(t_i) \,\square\, F(t_{i-1} - t_i) \,\square\, \ldots \,\square\, F(t_2 - t_3) \,\square\, G(t_1 - t_2)$$

on

$$S_i = \{(t_1, \ldots, t_i) : t_1 > t_2 > \ldots > t_i\}$$

(*) This hypothesis is only for simplifying the treatment; it is not difficult however, to extend the results to the case of locally square integrable functions.

The triplet F(t), G(t), H(t) is called the factorization, and m its dimension. A factorization F(t), G(t), H(t) is said to be minimal when its dimension assumes the smallest value over all the factorization of the sequence.

For the purpose of giving a condition of minimality, describing all (minimal and non minimal) factorizations and finding an algorithm for reducing a given factorization to its minimal form, we need introducing some matrices formed with F(t), G(t), H(t) and prove firstly a number of basic lemmas.

We start by introducing the sequences of matrices(*)

$$\bar{P}_1(t_1) = G(t_1)$$

$$\bar{P}_i(t_1, \ldots, t_i) = [F^{(1)}(t_i)\bar{P}_{i-1}(t_1, \ldots, t_{i-1}) \ldots$$

$$\ldots F^{(p)}(t_i)\bar{P}_{i-1}(t_1, \ldots, t_{i-1})] \qquad (i = 2, 3, \ldots) \qquad (A.3)$$

$$P_i[F,G](t_1, \ldots, t_i) = [\bar{P}_1\, \bar{P}_2 \ldots \bar{P}_i]$$

where $F^{(i)}(t)$ are mxm partitions of F(t) $(i = 1, 2, \ldots, p)$,
and the sequence

$$\bar{Q}_1(t_1) = \begin{bmatrix} H^{(1)}(t_1) \\ \vdots \\ H^{(p)}(t_1) \end{bmatrix}$$

$$\bar{Q}_i(t_1, \ldots, t_i) = \begin{bmatrix} \bar{Q}_{i-1}(t_1, \ldots, t_{i-1})F^{(1)}(t_i) \\ \vdots \\ \bar{Q}_{i-1}(t_1, \ldots, t_{i-1})F^{(p)}(t_i) \end{bmatrix} \qquad (i = 2, 3, \ldots) \qquad (A.4)$$

(*) The superscript (i) is merely a way of indexing the blocks; this choice is imposed by the need of further subscripts for denoting other partitions (see, for example, (A.21)).

$$Q_i[F, H](t_1, \ldots, t_i) = \begin{bmatrix} \bar{Q}_1 \\ \bar{Q}_2 \\ \cdot \\ \cdot \\ \cdot \\ \bar{Q}_i \end{bmatrix}$$

and the corresponding gramian matrices

(A.5) $P_i[F,G] = \int_{\Delta_t} P_i P_i^* \, dt_1 \cdots dt_i$

(A.6) $Q_i[F,H] = \int_{\Delta_t} Q_i^* Q_i \, dt_1 \cdots dt_i$

where Δ_i is the interval $\{(t_1, \ldots, t_i) : 0 \le t_k \le 1, \forall k\}$.

In connection with the above matrices it is possible to prove the following

Lemmas.

(A.7) **Lemma** — There exists an integer k such that

(A.8) $\mathcal{R}(\mathcal{P}_i) \subset \mathcal{R}(\mathcal{P}_{i+1})$ $\forall i < k$

(A.9) $\mathcal{R}(\mathcal{P}_i) \equiv \mathcal{R}(\mathcal{P}_{i+1})$ $\forall i \ge k$

and furthermore

(A.10) $k \le m$

Proof — (A.8) and (A.9) can be proved by first showing that

(A.11) $\mathcal{R}(\mathcal{P}_i) \subseteq \mathcal{R}(\mathcal{P}_{i+1})$ $\forall i$

In fact

(A.12) $\mathcal{P}_{i+1} = \mathcal{P}_i + \int_{\Delta_{i+1}} \bar{P}_{i+1} \bar{P}_{i+1}^T \, dt_1 \cdots dt_{i+1}$

Bearing in mind that the gramian matrices are non negative definite, it follows that for each vector $z \epsilon R^m$, $z \epsilon \mathcal{N}(\mathcal{P}_{i+1}) \Rightarrow z \epsilon \mathcal{N}(\mathcal{P}_i)$, i.e. $\mathcal{N}(\mathcal{P}_{i+1}) \subseteq \mathcal{N}(\mathcal{P}_i)$ and relation (A. 11) is therefore obtained by means of orthogonal complementation.

Moreover, it can be shown that

$$\mathcal{R}(\mathcal{P}_{i-1}) = \mathcal{R}(\mathcal{P}_i) \Rightarrow \mathcal{R}(\mathcal{P}_i) = \mathcal{R}(\mathcal{P}_{i+1})$$ (A.13)

In fact, it follows from the hypothesis in (A.13) that there exists a matrix R and a nonsingular matrix S, such that

$$\mathcal{P}_{i-1} = RR^*$$ (A.14′)

$$\mathcal{P}_i = RSS^*R^*$$ (A.14″)

Moreover

$$\mathcal{P}_i = \mathcal{P}_1 + \sum_{j=1}^{p} \int_0^1 F^{(j)}(t)\mathcal{P}_{i-1} F^{(j)^*}(t)dt$$ (A.15)

It follows from (A.14′) and (A.15) that

$$z \epsilon \mathcal{N}(\mathcal{P}_i) \Rightarrow z \epsilon \mathcal{N}(\mathcal{P}_1) \qquad \& \qquad z^* F^{(j)}(t) R \equiv 0 \quad \forall j$$ (A.16)

and, consequently, from (A.15) written for i+1 and (A.14″), one obtains

$$z \epsilon \mathcal{N}(\mathcal{P}_i) \Rightarrow z \epsilon \mathcal{N}(\mathcal{P}_{i+1})$$ (A.17)

Hence

$$\mathcal{R}(\mathcal{P}_i) \supseteq \mathcal{R}(\mathcal{P}_{i+1})$$ (A.18)

This, together with (A.11), proves (A.13).

The validity of (A.11) and (A.13) proves the existence of an integer k such that (A.8) and (A.9) are fulfilled.

As regards (A.10), it is sufficient to bear in mind that

(A.19) $\mathcal{R}(\mathcal{P}_i) \subseteq R^m$ $\forall i$

(A.20) Lemma — If only $\tilde{m} < m$ rows of $P_m [F,G] (t_1, \ldots, t_m)$ are linearly independent over Δ_m, then there exists a constant non-singular mxm matrix T such that

$$TF^{(i)}(t)T^{-1} = \begin{bmatrix} F_{11}^{(i)}(t) & F_{12}^{(i)}(t) \\ \\ 0 & F_{22}^{(i)}(t) \end{bmatrix} \quad i = 1, \ldots, p)$$

(A.21)

$$T \, G(t) = \begin{bmatrix} G_1(t) \\ \\ 0 \end{bmatrix}$$

where the matrices $F_{11}^{(i)}(t)$ and $G_1(t)$ are respectively $\tilde{m} \times \tilde{m}$ and $\tilde{m} \times q$, and $P_{\tilde{m}} [F_{11}, G_1](t_1, \ldots, t_{\tilde{m}})$, with

(A.22) $F_{11}(t) = [F_{11}^{(1)}(t) \ldots F_{11}^{(p)}(t)]$

has its $\tilde{m}$ rows linearly independent over $\Delta_{\tilde{m}}$.

Proof — If only $\tilde{m}$ rows of $P_m [F,G]$ are linearly independent over Δ_m, there exists a constant non-singular mxm matrix T such that

(A.23) $T P_m [F,G](t_1, \ldots, t_m) = \begin{bmatrix} \tilde{P}(t_1, \ldots, t_m) \\ \\ 0 \end{bmatrix}$

with the $\tilde{m}$ rows of $\tilde{P}(t_1, \ldots, t_m)$ linearly independent over Δ_m.

Bearing in mind the definition (A.3), (A.23) yields

(A.24) $T\bar{P}_1(t_1) = TG(t_1) = \begin{bmatrix} X_1(t_1) \\ \\ 0 \end{bmatrix}$

$$T\bar{P}_2(t_1,t_2) = [T\,F^{(1)}(t_2)G(t_1)\dots T\,F^{(p)}(t_2)\,G(t_1)] = \begin{bmatrix} X_2(t_1,t_2) \\ \\ 0 \end{bmatrix} \qquad (A.24)$$

where $X_1(t_1)$, $X_2(t_1,t_2)$, $\dots$ are used to denote block columns of $\tilde{P}(t_1,\dots,t_m)$. The first equation of (A.24) coincides with the last of (A.21) if one puts $X_1(t) = = G_1(t)$. From the second equation of (A.24) introducing the partition

$$TF^{(i)}(t)\,T^{-1} = \begin{bmatrix} F_{11}^{(i)}(t) & F_{12}^{(i)}(t) \\ \\ F_{21}^{(i)}(t) & F_{22}^{(i)}(t) \end{bmatrix} \qquad (A.25)$$

one obtains

$$\begin{bmatrix} X_2(t_1,t_2) \\ \\ 0 \end{bmatrix} = \left[\begin{pmatrix} F_{11}^{(1)}(t) & F_{12}^{(1)}(t) \\ \\ F_{21}^{(1)}(t) & F_{22}^{(1)}(t) \end{pmatrix} \begin{pmatrix} G_1(t_1) \\ \\ 0 \end{pmatrix} \dots \right.$$

$$(A.26)$$

$$\left. \dots \begin{pmatrix} F_{11}^{(p)}(t) & F_{12}^{(p)}(t) \\ \\ F_{21}^{(p)}(t) & F_{22}^{(p)}(t) \end{pmatrix} \begin{pmatrix} G_1(t_1) \\ \\ 0 \end{pmatrix} \right] = \begin{bmatrix} F_{11}^{(1)}G_1(t_1) & F_{11}^{(p)}G_1(t_1) \\ & \dots & \\ F_{21}^{(1)}G_1(t_1) & F_{21}^{(p)}G_1(t_1) \end{bmatrix} .$$

Proceeding in a similar way with the remaining elements of (A.25), and bearing in mind the results of type (A.26) that will gradually be acquired, one obtains

$$\tilde{P}(t_1,\dots,t_m) = P_m[F_{11},G_1](t_1,\dots,t_m) \qquad (A.27)$$

with $F_{11}(t)$ given by (A.22); therefore, by hypothesis, the $\tilde{m}$ rows of the r.h.s. of

(A.27) are linearly independent over Δ_m and this implies, by Lemma (A.7), that the $\widetilde{m}$ rows of $P_{\widetilde{m}}[F_{11}, G_1](t_1, \ldots, t_{\widetilde{m}})$ are linearly independent over $\Delta_{\widetilde{m}}$. Moreover, from (A.26) one also obtains

(A.28)
$$0 = [F_{21}^{(1)}(t)P_{m-1}[F_{11}, G_1](t_1, \ldots, t_{m-1}) \cdots$$
$$\cdots F_{21}^{(p)}(t)P_{m-1}[F_{11}, G_1](t_1, \ldots, t_{m-1})]$$

Since, as proved, above, the $\widetilde{m}$ rows of $P_{\widetilde{m}}[F_{11}, G_1](t_1, \ldots, t_{\widetilde{m}})$ are linearly independent over $\Delta_{\widetilde{m}}$ and $\widetilde{m} < m$, also the $\widetilde{m}$ rows of $P_{m-1}[F_{11}, G_1](t_1, \ldots, t_{m-1})$ are linearly independent over Δ_{m-1} and this, by (A.28) implies $F_{21}^{(i)}(t) \equiv 0$, $i = 1, \ldots, p$. This completes the proof.

Similar Lemmas can be proved for matrices Q_i and $\mathcal{Q}_i$.

(A.29) Lemma — There exists an integer k such that

(A.30) $$\mathcal{R}(\mathcal{Q}_i) \subset \mathcal{R}(\mathcal{Q}_{i+1}) \qquad \forall i < k$$

(A.31) $$\mathcal{R}(\mathcal{Q}_i) \equiv \mathcal{R}(\mathcal{Q}_{i+1}) \qquad \forall i \geq k$$
and, furthermore

(A.32) $$k \leq m$$

(A.33) Lemma — If only $\widetilde{m} < m$ columns of $Q_m[F, H](t_1, \ldots, t_m)$ are linearly independent over Δ_m, then there exists a constant non-singular mxm matrix T such that

(A.34)
$$TF^{(i)}(t)T^{-1} = \begin{bmatrix} F_{11}^{(i)}(t) & 0 \\ F_{21}^{(i)}(t) & F_{22}^{(i)}(t) \end{bmatrix}$$

$$H^{(i)}(t)T^{-1} = [H_1^{(i)}(t) \quad 0] \qquad (i = 1, \ldots, p)$$

where the matrices $F_{11}^{(i)}(t)$ and $H_1^{(i)}(t)$ are respectively $\tilde{m} \times \tilde{m}$ and $q \times \tilde{m}$ and $Q_{\tilde{m}}[F_{11},$ $,H_1](t_1, \ldots, t_{\tilde{m}})$, with $F_{11}(t)$ given by (A.22) and

$$H_1(t) = [H_1^{(1)}(t) \ldots H_1^{(i)}(t)] \qquad (A.35)$$

has its $\tilde{m}$ columns linearly independent over $\Delta_{\tilde{m}}$.

Finally, based on the Definition (A.1), we can establish a property characterizing all the factorizations of a given sequence of kernels.

(A.36) Lemma — For each fixed r and s, the product $Q_r[F,H] \cdot P_s[F,G]$ is invariant and the set of all factorization of a given factorizable sequence of kernels.

Proof — This consist in showing that the elements of the product under considerations coincide with the elements of the kernels of the sequence, calculated for appropriate arguments and suitably ordered. In fact, performing the product $Q_r[F,H](t_1, \ldots, t_r) \, P_s[F,G](t_{r+1}, \ldots, t_{r+s})$, one obtains a pattern of this type (for simplicity, the argument are omitted)

$$(A.37) \quad \begin{pmatrix} \begin{pmatrix} H^{(1)}G \\ \vdots \\ H^{(p)}G \end{pmatrix} & \begin{pmatrix} H^{(1)}F^{(1)}G \ldots H^{(1)}F^{(p)}G \\ \vdots \qquad\qquad \vdots \\ H^{(p)}F^{(1)}G \ldots H^{(p)}F^{(p)}G \end{pmatrix} & \begin{matrix} \text{blocks} \\ \text{of} \\ W_4 \end{matrix} \\ \begin{pmatrix} H^{(1)}F^{(1)}G \\ \vdots \\ H^{(p)}F^{(1)}G \\ \vdots \\ H^{(1)}F^{(p)}G \\ \vdots \\ H^{(p)}F^{(p)}G \end{pmatrix} & \begin{matrix} \text{blocks} \\ \text{of} \\ W_4 \end{matrix} & \ldots \\ \text{blocks of } W_4 & \ldots & \ldots \end{pmatrix}$$

Now we readily recognize that the blocks of the superblock in the position (1.1) are the blocks of a suitable partition of W_2 (in fact, $W_2 = (H^{(1)}G \ldots H^{(p)}G)$); the same is true for the super-blocks (1.2) and (2.1) with regard to W_3, and so on. The superblock in the position (i, j) contains all the blocks (and nothing more) of a suitable partition of W_{i+j}, calculated for appropriate arguments.

We are now in the position of proving the previously mentioned results on the factorizable sequences. The first one concerns minimality and is expressed by the following :

(A.38) Theorem — An m-dimensional factorization $\{F(t), G(t), H(t)\}$ of a factoriza-ble sequence of kernels is minimal if and only if the rows of $P_m\,[F,G]$ and the col-umns of $Q_m\,[F,G]$ are linearly independent over Δ_m. The minimal factorizations are an equivalence class with respect to the relation

$$(A.39) \quad \{F_1(t), G_1(t), H_1(t)\} \sim \{F_2(t), G_2(t), H_2(t)\} \Leftrightarrow \begin{cases} F_1^{(i)}(t) = TF_2^{(i)}(t)T^{-1} \\ G_1(t) = TG_2(t) \\ H_1^{(i)}(t) = H_2^{(i)}(t)T^{-1} \\ \qquad (i = 1, \ldots, p) \end{cases}$$

Proof — The proof will be divided into two parts, the first relating to the conditions of minimality and the second to the conditions of equivalence.

1st Part : Necessity — Suppose, for example that the matrix $P_m\,[F,G]$, corresponding to a minimal factorization, has only $\tilde{m} < m$ linearly independent rows. As a conse-quence of Lemma (A.20), there will then exist a constant non-singular mxm matrix T such that

$$(A.40) \quad T\,F^{(i)}(t)T^{-1} = \begin{bmatrix} F_{11}^{(i)}(t) & F_{12}^{(i)}(t) \\ 0 & F_{22}^{(i)}(t) \end{bmatrix} \qquad T\,G(t) = \begin{bmatrix} G_1(t) \\ 0 \end{bmatrix}$$

for $i = 1, \ldots, p$. If one now takes into consideration also the partition

$$H^{(i)}(t)T^{-1} = (H_1^{(i)}(t) \ H_2^{(i)}(t)) \tag{A.41}$$

one can immediately see that the triplet $\{F_{11}(t), G_1(t), H_1(t)\}$, with $F_{11}(t)$ and $H_1(t)$ defined as in (A.22) and (A.35), which has a dimension $\tilde{m} < m$, is also a factorization of the sequence $\{W_i(t_1, \ldots, t_i)\}_2^{\infty}$ and this contradicts the hypothesis of minimality. A similar proof, based on Lemma (A.33), can be given for the matrix $Q_m \{F, H\}$.

1st Part : Sufficiency — Suppose that the factorization $\{F(t), G(t), H(t)\}$, for which the matrices $P_m [F, G]$ and $Q_m [F, H]$ have, respectively, their rows and their columns linearly independent, is not a minimal one and that, consequently, there exists a factorization $\{\tilde{F}(t), \tilde{G}(t), \tilde{H}(t)\}$ of dimension $\tilde{m} < m$ for the same sequence $\{W_i(t_1, \ldots, t_i)\}_2^{\infty}$. Considering now the matrices $P_m [\tilde{F}, \tilde{G}]$ and $Q_m [\tilde{F}, \tilde{H}]$, by Lemma (A.36) we have

$$Q_m [F, H] (t_1, \ldots, t_m) P_m [F, G] (t_{m+1}, \ldots, t_{2m}) =$$
$$= Q_m [\tilde{F}, \tilde{H}] (t_1, \ldots, t_m) P_m [\tilde{F}, \tilde{G}] (t_{m+1}, \ldots, t_{2m}). \tag{A.42}$$

Premultiplying both sides of (A.42) by $Q_m^* [F, H](t_1, \ldots, t_m)$, postmultiplying by $P_m^* [F, G](t_1, \ldots, t_m)$ and then integrating one obtains

$$\mathcal{Q}_m \cdot \mathcal{P}_m = \int_{\Delta_m} Q_m^* [F, H] \, Q_m [\tilde{F}, \tilde{H}] dt_1 \ldots dt_m \cdot$$

$$\cdot \int_{\Delta_m} P_m [\tilde{F}, \tilde{G}] P_m^* [F, G] \, dt_{m+1} \ldots dt_{2m}. \tag{A.43}$$

The rank of the matrix on the l.h.s. is equal to m by hypothesis, while the rank of the matrix on the r.h.s. does not exceed $\tilde{m}$; it follows that

(A.44) $$m \leq \tilde{m}.$$

This contradicts the hypothesis that $\tilde{m} < m$ and completes the proof of the first part.

2nd Part — If the triplet $\{F(t), G(t), H(t)\}$ is a minimal factorization of the given sequence of kernels, it can be verified immediately that the triplet $\{TF(t)T^{-1}, TG(t), H(t)T^{-1}\}$ is also a minimal factorization (see (A.2)). Viceversa, let $\{F_1(t), G_1(t), H_1(t)\}$ and $\{F_2(t), G_2(t), H_2(t)\}$ be any two minimal factorizations of the given sequence of kernels. Considering the matrices $Q_m^{\cdot}[F_1,H_1]$, $P_m[F_1,G_1]$, $Q_m[F_2, H_2]$ and $P_m[F_2,G_2]$, one finds (see Lemma (A.36)) that

(A.45)
$$Q_m[F_1,H_1](t_1,\ldots,t_m) \cdot P_m[F_1,G_1](t_{m+1},\ldots,t_{2m}) =$$
$$= Q_m[F_2,G_2](t_1,\ldots,t_m) \cdot P_m[F_2,G_2](t_{m+1},\ldots,t_{2m}).$$

By virtue of the results proved in the first part, the m columns of $Q_m[F_1,H_1]$ and $Q_m[F_2,H_2]$ and the m rows of $P_m[F_1,G_1]$ and $P_m[F_2,G_2]$ are linearly independent over Δ_m. Consequently, generalizing the methods and results established by D.C. Youla in [10] with regard to functions defined in R^1, one finds that

(A.46)
$$Q_m[F_1, H_1] = Q_m[F_2, H_2] T^{-1}$$
$$P_m[F_1, G_1] = T P_m[F_2, G_2]$$

where T is a constant non-singular matrix. Bearing in mind the expressions (A.3) and (A.4), it follows that

(A.47) $$H_1^{(i)}(t) = H_2^{(i)}(t)T^{-1}$$
$$(i = 1,\ldots, p)$$

(A.48) $$G_1(t) = TG_2(t)$$

As regards the relation between $F_1(t)$ and $F_2(t)$, one can first of all write (see Lemma (A.36))

$$Q_m[F_1,H_1](t_1,\ldots,t_m) \cdot F_1^{(i)}(t_{m+1}) \cdot P_m[F_1,G_1](t_{m+2},\ldots,t_{2m+1}) =$$

$$= Q_m[F_2,H_2](t_1,\ldots,t_m) \cdot F_2^{(i)}(t_{m+1}) \cdot \tag{A.49}$$

$$\cdot P_m[F_2,G_2](t_{m+2},\ldots,t_{2m+1})$$

for all i. Substituting the expressions (A.46) in the l.h.s. of (A.49) and bearing in mind the linear independence of the columns of $Q_m[F_2,H_2]$ and of the rows of $P_m[F_2,G_2]$, one finally obtains

$$F_1^{(i)}(t) = T F_2^{(i)}(t) T^{-1} \qquad (i=1,\ldots,p) \tag{A.50}$$

The expressions (A.47), (A.48) and (A.50) complete the proof of the second part.

Together with the description of all the minimal factorizations it is convenient to find an expression for all the possible factorizations. In this connection one can state the following

(A.51) Theorem − Any factorization of a factorizable sequence of kernels can be written in the form

$$F^{(i)}(t) = T \begin{bmatrix} F_0^{(i)}(t) & 0 & F_{13}^{(i)}(t) & F_{14}^{(i)}(t) \\ F_{21}^{(i)}(t) & F_{22}^{(i)}(t) & F_{23}^{(i)}(t) & F_{24}^{(i)}(t) \\ 0 & 0 & F_{33}^{(i)}(t) & F_{34}^{(i)}(t) \\ 0 & 0 & F_{43}^{(i)}(t) & F_{44}^{(i)}(t) \end{bmatrix} T^{-1} \tag{A.52}$$

$$G(t) = T \begin{bmatrix} G_0(t) \\ G_2(t) \\ 0 \\ 0 \end{bmatrix} \qquad H^{(i)}(t) = (H_0^{(i)}(t) \quad 0 \quad H_3^{(i)}(t) \quad H_4^{(i)}(t)) T^{-1}$$

where T is a constant non-singular matrix, and the triplet $\{F_o(t), G_o(t), H_o(t)\}$, with

(A.53) $F_o(t) = [F_o^{(i)}(t) \ldots F_o^{(p)}(t)]$

(A.54) $H_o(t) = [H_o^{(1)}(t) \ldots H_o^{(p)}(t)]$

is a minimal factorization.

Proof – If $\{F(t), G(t), H(t)\}$ is a non minimal factorization of the sequence $\{W_i(t_1, \ldots, t_i)\}_2^\infty$, it is possible to apply a reduction process that makes successive use of Lemmas (A.20) and (A.33) as indicated in the proof of the 1st Part (necessity) of Theorem (A.38). Denoting the number of linearly independent rows of $P_m[F,G]$ by m_1, it follows from Lemma (A.20) that there exists an mxm matrix T_1 such that one can write

(A.55)
$$T_1 F^{(i)}(t) T_1^{-1} = \begin{bmatrix} F_a^{(i)}(t) & F_{ab}^{(i)}(t) \\ 0 & F_b^{(i)}(t) \end{bmatrix} \qquad T_1 G(t) = \begin{bmatrix} G_a(t) \\ 0 \end{bmatrix}$$

$$H^{(i)}(t) T_1^{-1} = (H_a^{(i)}(t) \ H_b^{(i)}(t))$$

for all i, and that the matrix $P_{m_1}[F_a, G_a]$ has linearly independent rows. If m_2 is now used to denote the number of linearly independent columns of $Q_{m_1}[F_a, H_a]$, it follows from Lemma (A.33) that there exists an $m_1 x m_1$ matrix T_2 such that one can write

(A.56a)
$$T_2 F_a^{(i)}(t) T_2^{-1} = \begin{bmatrix} F_c^{(i)}(t) & 0 \\ & F_d^{(i)}(t) \end{bmatrix} \qquad T_2 G_a(t) = \begin{bmatrix} G_c(t) \\ G_d(t) \end{bmatrix}$$

$$H_a^{(i)}(t)T_2^{-1} = (H_c^{(i)}(t) \quad 0) \qquad (A.56b)$$

and that the matrix $Q_{m_2}[F_c, H_c]$ has linearly independent columns. Furthermore, it can be verified that the rows of $P_{m_2}[F_c, G_c]$ are also linearly independent by construction. Since it follows from the structure of (A.55) and (A.56) that the triplet $\{F_c(t), G_c(t), H_c(t)\}$ is also a factorization of the given sequence, then, by virtue of Theorem (A.38), the above conditions imply that this triplet is a minimal factorization. The expressions (A.52) follow from this fact if one chooses

$$T = T_2^{-1}\begin{bmatrix} T_2^{-1} & 0 \\ 0 & I \end{bmatrix} \qquad (A.57)$$

(A.58) **Remark** – Only the minimal part of the factorization has been enlighted in stating and proving Theorem (A.51); however, also the typical pattern of the canonical forms of the linear cases could have been made to appear.

(A.59) **Corollary** – The dimension m_o of a minimal factorization of a factorizable sequence of kernels is given by

$$m_o = \text{rank} \{\mathcal{Q}_m[F,H]\mathcal{P}_m[F,G]\} \qquad (A.60)$$

where $\{F(t), G(t), H(t)\}$ is any given factorization of the same sequence.

Proof – Replacing $\{F(t), G(t), H(t)\}$ with the expressions obtained from (A.52) and then calculating the r.h.s. of (A.60), it is easy to see that

$$\text{rank} \{\mathcal{Q}_m[F,H]\mathcal{P}_m[F,G]\} = \text{rank} \{\mathcal{Q}_m[F_o,H_o]\mathcal{P}_m[F_o,G_o]\} \qquad (A.61)$$

and (A.60) follows directly from this.

REFERENCES

[1]   P. d'Alessandro, A. Isidori and A. Ruberti, **Realization and structure Theory of Bilinear Dynamical Systems**. Reports 2.04 and 2.16, Istituto di Automatica, University of Rome (1972), submitted for publication. Part of the results on the Structure Theory were referred in the paper : **Structure Analysis of Linear and Bilinear Dynamical Systems**, on "Theory and Applications of Variable Structure Systems" (R.R. Mohler and A. Ruberti, eds.), Academic Press New York (1972), pp. 25 - 36.

[2]   C. Bruni, G. Di Pillo and G. Koch, **On the Mathematical Models of Bilinear Systems**, Ricerche di Automatica, 2 (1971), pp. 11 - 26.

[3]   L. Zadeh and C.A. Desoer, "Linear System Theory", McGraw Hill, New York (1963).

[4]   R.W. Brockett, **System Theory on Group Manifold and Coset Spaces**, SIAM J. on Control, 10 (1972), pp. 265 - 284.

[5]   R.W. Brockett, **On the Algebraic Structure of Bilinear Systems**, on "Theory and Applications of Variable Structure Systems" (R.R. Mohler and A. Ruberti, eds.), Academic Press, New York (1972), pp. 153 - 168.

[6]   R.E. Rink and R.R. Mohler, **Completely Controllable Bilinear Systems** SIAM J. on Control, 6 (1968), pp. 477 - 486.

[7]   T.J. Tarn, D.L. Elliot and T. Goka, **Controllability of Discrete Bilinear Systems with Bounded Control**, IEEE Trans. on Automatic Control, 18 (1973), to appear.

[8]   P. d'Alessandro, **Structural Properties Invariance and Insensitivity of Discrete-Time Bilinear Systems**, Ricerche di Automatica, 3 (1972) pp. 158 - 169.

[9]   G. Basile, and G. Marro, Luoghi Caratteristici dello Spazio degli Stati relativi al Controllo dei Sistemi Lineari, L'Elettrotecnica, 55 (1968), pp. 846 - 852.

[10]   A. Isidori, Direct Construction of Minimal Bilinear Realizations from Non-linear Input/Output Maps, Report 3.02, Istituto di Automatica, University of Rome (1973), submitted for publication.

[11]   B.L. Ho and R.E. Kalman, Effective Construction of Linear State Variable Models from Input/Output Functions, Regelungstecknik, 14 (1966), pp. 545 - 548.

[12]   C. Gori-Giorgi and A. Isidori, A Simple Algorithm for Factorizing the Weighting Pattern, Ricerche di Automatica, 2 (1971), pp. 288 - 294.

[13]   G. Koch, A Realization Theorem for Infinite Dimensional Bilinear Systems, Ricerche di Automatica, 3 (1972), to appear.

[14]   D.C. Youla, The Synthesis of Linear Dynamical Systems from Prescribed Weighting-Pattern, SIAM J. on Appl. Mathe., 14 (1966), pp. 527 - 549.

CONTENTS